RC
O·

AMERICAN NORMAL

AMERICAN NORMAL

THE HIDDEN WORLD
OF ASPERGER SYNDROME

Lawrence Osborne

COPERNICUS BOOKS

An Imprint of Springer-Verlag

Additional copyrighted material is cited in the Acknowledgments, on page 214, which constitutes an extension of this copyright page.

Published in the United States by Copernicus Books,
an imprint of Springer-Verlag New York, Inc.
A member of BertelsmannSpringer Science+Business Media GmbH

Copernicus Books
37 East 7th Street
New York, NY 10003
www.copernicusbooks.com

Library of Congress Cataloging-in-Publication Data
Osborne, Lawrence
    American normal: the hidden world of Asperger syndrome / Lawrence Osborne.
        p.   cm.
    Includes bibliographical references and index.
    ISBN 0-387-95307-8
        1. Asperger syndrome.   I. Title.
    RC553.A88 O83   2002
    616.89'82—dc21   2002073782

Manufactured in the United States of America.
Printed on acid-free paper.

9 8 7 6 5 4 3 2 1

ISBN 0-387-95307-8   SPIN 10839778

CONTENTS

PREFACE

Hardly a month goes by without a new article or a television special about Asperger Syndrome. We are told that it's the fastest-growing psychiatric condition among the children of Silicon Valley (where it's been called "the geek syndrome"); that tens of thousands of children and adults suffer from it; and that the lofty Thomas Jefferson probably had it, too. If people have heard of one psychiatric condition other than schizophrenia, it's often the subtlest form of autism known as Asperger Syndrome, a sometimes poetic deformation of the mind that turns people into solitary misfits but that also makes them virtuosos—sometimes of valued skills, but often of madly irrelevant obsessions. Asperger Syndrome has become unexpectedly fashionable. More, it has become perhaps the first *desirable* syndrome of the twenty-first century: a terrible burden, yes, but also a proof of eccentric intelligence, of genius, even—and at the very least, of that increasingly rare commodity, individuality.

Even the tiniest Asperger boy can often be a walking *Encylopedia Britannica* entry on different species of cicadas, obscure clock-manufacturing companies, telephone cable insulating firms, the passenger list of the *Titanic*, baseball trivia from 1921, say, to 1922, or the provincial capitals of Brazil. In one documented case, a child memorized the addresses, telephone numbers, and zip codes of every member of Congress, while one family of a Long Island Asperger child had to make continual diversions in order to visit the site of the TWA airline disaster. Another three-year-old Asperger boy disassembled and reassembled a refrigerator motor. Very amusing. But Asperger Syndrome also afflicts me with a distinct unease: it remains so difficult so diagnose, so restlessly vague. And, worse still, it strikes too close to home.

I find that, when I look at the matter closely, I am sure (or at least half sure) that the symptoms of this now *à la mode* syndrome are already familiar to me. Admittedly, I knew nothing when I was a child about deep-fat friers, and even then I was stonily indifferent to the phone numbers of politicians. But I did know all the tank designs used by General Guderian on the Eastern Front in 1941, and at the age of nine I knew by heart the complicated sex lives of all the characters in *The Odyssey*. Was I mad?

As it happens, in the psychiatric Tower of Babel which America is fast becoming, perhaps all of us can consider the case of the Asperger loner with more than a slight misgiving. Who among us does not have strange little compulsions and obsessions, which, while not exactly awarding us membership in the Asperger's club, certainly must give us pause about our normality? Yet there is, of course, a great difficulty in writing about psychiatry. I am not a psychiatrist and neither, most probably, are you. We wonder to ourselves if we even have the right to entertain opinions concerning this intricate, not to say mystifying, science, if indeed it is a science. But here's the rub. The object of this science is ourselves and our normality. That is to say, our basic nature. Do we actually possess our normality, or (a distinctly less attractive possibility) does *it* in fact possess *us*? About this, we can hardly fail to hold a view.

I would probably never have set out on a journey into the topsy-turvy Land of Asperger's if I hadn't been tormented by this ominous question, which of course has no obvious answer. All that exists is the journey itself, which was conducted in the spirit of a merrily admitted ignorance. After all, the experts have already spoken on this curious subject, and what they have to say is widely available. What I wanted on this voyage, this road trip, was to visit the places and people hidden, often in plain sight: not just experts, or caregivers, or even just children, but Asperger people (if they can be called that) living out their very lives. What I wanted, in short, was merely to let a journey speak for itself. The characters met on the way, after all, were not psychiatric oddities, dark goblins inhabiting the infamous *Diagnostic Manual*, but only varyingly intense and wayward variations of myself.

INTRODUCTION

The maps of Asperger Syndrome have been drawn and redrawn over fifty years, but the borders remain maddeningly vague.

The disorder, sometimes called a form of "high-functioning autism," was first pointed out by, then named after the Viennese pediatrician Hans Asperger in 1944. The phrase "high-functioning" is meant to distinguish Asperger's from classical autism—the latter condition is typically characterized by much more obvious deficits in speech, intelligence, and development. Asperger's sufferers, in contrast, appear largely normal. Or almost normal. They can function intellectually at a high level and can, more or less, blend into the general population. Nonetheless, whether Asperger's is or is not on a continuum with autism (the issue is not resolved), it most assuredly can be what the distinguished researcher and writer Uta Frith has called "a devastating handicap." In the United States, the syndrome was only made "official" with its entry into the fourth edition of the American Psychiatric Association's *Diagnostic and Statistical Manual of Mental Disorders* (DSM-IV) in 1994.

What, then, is a reasonable definition? Perhaps the most workable one I've encountered is from researchers and writers at the Yale Child Study Center:

> Paucity of empathy; naive, inappropriate, one-sided social interaction, little ability to form friendships and consequent social isolation; pedantic and monotonic speech; poor nonverbal communication; intense absorption in circumscribed topics such as the weather, facts about TV stations, railway timetables or maps, which are learned by rote fashion and reflect poor understanding, conveying the impression of eccentricity; and clumsy and ill-coordinated movements and odd posture.

The most workable, but still unsatisfying. Through these clumsy and ill-coordinated clouds of psychiatric prose, one glimpses a unique condition. Asperger people are not idiot savants like Rain Man or head-banging mental patients rocking in their chairs and screaming;

they do not conform in any way to the clichés roused in us by the word "autistic." Instead, the cognitive disability appears to be purely, or almost purely, social. Essentially, for reasons that are completely unknown, Asperger people cannot read the human face or its emotions. They cannot learn social rules, nuances, or metaphors. Often brilliant intellectually, they cannot read the simplest social cue or hint: instead, rigid obsessions, often numerical, dominate their inner life. And they live with the affliction for the whole of their lives.

★

It's a curious fact that a great many people in the U.S. who have Asperger Syndrome are self-diagnosed. As the number of people designated as being on the autistic spectrum rises, I have the feeling that thousands of people like myself are re-examining their childhoods with a certain anxiety, but not without a certain relish as well. Every tic they have ever had is now suspect, a sign of something systemic and previously concealed. A century of widespread psychology and psychologizing has made this apprehensive mind-set respectable. Eccentricity itself is less and less accepted as an innocent aberration, a potentially *fruitful* quirk of character, for the question of normality imposes itself constantly. *Did* you play the lute when you were a boy, or not? *Did* you line up your toys in rows or spin on your heels imitating a propeller for hours on end? In a culture defined by obsessive navel-gazing, we have taken to using our navels as medical crystal balls. What disorder do we have? What form of autism do we think we have, however slight and superficial? And, most importantly, which section or subsection of the *Diagnostic Manual* do we fit into?

Characteristically, people often now describe themselves as having "Aspergerish traits" without actually going so far as to call themselves autistic. Having Aspergerish traits is today one of the most fashionable self-diagnoses in America, while autism is still a dread word. For Asperger people have a reputation for cleverness, subtlety, and even for genius. Einstein is now frequently claimed as an Asperger's genius, as are the pianist Glenn Gould and the composer Béla Bartók. In 1996, *Time* magazine even ran a piece entitled "Diagnosing Bill Gates," in which the nabob of Microsoft was roundly defined as a classic Asperger's type. If the richest man in the world has Asperger's, why not you? Asperger Syndrome is indeed, as autism researcher Uta Frith puts it, "the first plausible variant to crystallize out of the autism spectrum"—and perhaps only the first of many. But where should we place the emphasis—on "variant" or "autism"? Clearly, Asperger's stands apart from autism in general, and it is no wonder the parents of so many brilliant middle-class Asperger boys grow abusive at the very mention of the word "autistic." For them, Asperger's is an asset, not what the Greeks called a fate.

★

For years, psychologists argued over whether the mental derailments observable in autistic children occurred because of disturbed parenting and a hostile environment, or because of in-built neurological disorders. The great psychoanalyst Bruno Bettelheim was the most noted proponent of the former idea and is accordingly reviled by parental activist groups and especially Asperger's support cells all over America. Indeed, even so much as mention the word "Bettelheim" at conferences and seminars devoted to Asperger's and you will immediately hear a murmur of scandalized disapproval.

In latter decades, the biological model has come to triumph in the domains of professional expertise, especially after the publication of Bernard Rimland's work on autism in the 1960s. In fact, not only Asperger Syndrome but virtually every developmental disorder is now seen as biological and genetic in origin. As Arthur Kleinman of the Harvard Medical School has written, "Biology has cachet with psychiatrists."

The vast and thorny ensuing debate cannot really be explored here, but it's apposite to remember that nothing is as simple as it looks. Richard DeGrandpre, author of *Ritalin Nation*, makes this comment about Attention Deficit Disorder, another affliction that is increasingly explained in terms of biology:

> More than anything, ADD represents a growing prejudice in our culture—led in large part by the powerful influence of psychiatry professionals and pharmaceutical companies—which is that personality and behavioral traits are inborn and biological.

The debate between the two camps seesaws endlessly back and forth, without any decisive outcome. People like DeGrandpre argue that ADD, for example, is largely a culturally constructed disorder, not the biological deficit so many drug-wielding psychiatrists like to claim. In other words, the speed-intoxicated culture itself induces attention-deficits (what DeGrandpre calls "pseudo- ADD") then fails to understand its own handiwork. How else can one explain the fact that levels of ADD are 20 times higher in the drug-prescribing U.S. than in Western Europe? It comes as some surprise, moreover, to discover that there is little in the way of hard scientific proof for a biological origin of many developmental disorders. But it matters little. With their gleaming promises that whatever is biological in origin can be manipulated by medicine and technology for the better of all, materialist and determinist models of the human mind are in tune with our age. Most harried doctors might say that whatever works, works. How many suicides, they ask, have been avoided through a judicious use of Prozac? The question is grimly compelling.

A biological conception of mental disorders, though, does not necessarily guarantee sweeping promises of cures—and Hans Asperger himself never suggested that a cure for "his" condition would eventually be found. Notions that there must be a cure, at least somewhere in the future, have crept into the deepest crevices of the American psyche: the premise that the soul is chemical in nature and can consequently be altered by chemical engineers, that the problems of happiness and social adjustment can be solved mechanistically by a brave new pharmacopoeia, and that no one is doomed to disorders or even to unpredictable moods unless it be because of medical malpractice, poverty, or an intolerance to drugs.

Lawrence Diller, in a popular work on current pediatric psychiatry called *Running on Ritalin*, argues that the vague but complex notion of "personality" has been abandoned by American psychiatry in favor of a neuro-chemical vision of the individual, in which different parts of the brain determine behaviors and moods. Psychotropic drugs intervene at these sites in order to remake the troubled person—end of story. It is what Peter Kramer in *Listening to Prozac* famously dubbed "cosmetic psycho-pharmacology." And although these cosmetic drugs have been criticized frequently in recent years, their lure remains powerful.

★

Doctors rarely admit how little they know about the workings of drugs like "serotonin uptake inhibitors," while pointing pragmatically instead to the sometimes-dramatic improvements they seem to produce in, say, chronic depressives. It's difficult to say where addiction and cure begin and end; and with children the question is even more obscure. But it is most often there, in childhood, where lifelong diagnoses are applied.

Child psychiatrists who rely heavily on prescription-writing often defend themselves by claiming that their treatments save families from uncontrollable forces in the disturbed boy or girl, which they doubtless sometimes do. But the means they employ are hardly those of an exact science, whatever the impressive-sounding vocabularies they employ. And it is precisely the vocabulary of psychiatry that is striking to me. Do we, the ignorant laymen, have any right to feel put off by it? Can we doubt the workings of treatments that seem to drag people from the edge of despair?

A recent exchange of views in the pages of *Salon* magazine between Lawrence Diller and Ross Greene, a psychologist at the Harvard Medical School, became snappish after Diller criticized Greene's book *The Explosive Child.* The latter recommends the use of psychotropic drugs to control children's behavior. Greene reacted angrily to Diller's criticisms, but also drew an explicit link between Asperger's and what he calls "non-compliant behavior." Adopting psychiatry's current language of behavioral management, Greene inveighed:

> . . . explosive/noncompliant children lack important skills related to managing frustration and handling demands for flexibility and adaptability. The goal of intervention flowing from this conceptualization is to teach these skills. Not by cajoling, but by having adults engage the child in a process by which important problem solving and conflict resolution skills—thinking of good solutions, anticipating problems before they arise, taking others' needs into account—are taught.

The child, in other words, is like a poorly performing junior executive. He or she has to learn "management skills," cooperation with the team, productive negotiating strategies. In the same vein, Greene continues:

> Dr. Diller also writes that *The Explosive Child* "overpathologizes" difficult children. Perhaps difficult children are more complicated than Dr. Diller is aware. Our research at Massachusetts General Hospital shows that noncompliant children almost always meet criteria for at least one other psychiatric condition, including attention deficit/hyperactivity disorder (ADHD), depression, bipolar disorder, anxiety disorders, nonverbal learning disability, language processing disorders, Tourette's disorder and Asperger's disorder. Our research at Mass. General also documents that the approach described in *The Explosive Child* is highly effective at reducing explosive outbursts, reducing adult-child conflict and, yes, improving a child's compliance.

Beyond the understandable self-interest of doctors advancing their own treatments, in this case a child-control system, one has to ask here if the egg comes before the chicken. Are children really a seething mass of pathological abnormalities, or have we made them into such? Greene suggests that every child is a candidate "for at least one other psychiatric condition"; but of course they are often candidates for most or even all of them. And the goal of bringing up children presumably should be the instilling of intimacy and respect, not "compliance" and legalized drug addiction. Diller's appeal for common sense, meanwhile, is aloofly dismissed as "unsophisticated." But it's an open question (as Diller himself asks it) what exactly the sophistication of the Harvard Medical School is all about. In short, profound questions in the psychiatric treatment of children remain entirely unresolved.

★

Some rebel pediatricians have begun to go much further in their defiance of what they see as a kind of psychiatric fundamentalism based on the *Diagnostic Manual*. Dr. Mel Levine is a Professor of Pediatrics at the University of North Carolina at Chapel Hill and the author of the recent book *A Mind at a Time*. As a developmental pediatrician working in schools, Levine has become increasingly dismayed at the way children are shoehorned into dubious categories of so-called Disorders.

"This whole thing," he said to me, "has become a huge problem in America. And it's not being subjected to any skeptical debate. We're pathologizing all human behavior, and in so doing, we're creating an institutionalized nightmare—a truly mad system in which everyone is 'sick.' The *Diagnostic Manual* is an absurd document, though of course it makes the American Psychiatric Association improbable amounts of money!" I asked him about Asperger's. "I for one am strongly opposed to the whole concept of Asperger Syndrome. It's yet another label around which the psychiatric industry can spin its usual paraphernalia. As for other would-be syndromes, I treat them with a high degree of skepticism. I refuse to even use the term ADHD in my clinic—I think it's monstrous. Children cannot be crushed by these reductionist labels." He laughed bitterly, or so I thought. It could have been merely ruefully. "I've banned all the D's from my practice!"

Levine thinks that American psychiatry embodies a deeply pessimistic, gloomily simplistic view of the world. Unable to conceive of a healthy eccentricity (or a truly complex individuality), it has elaborated a vast coding system instead. Every patient is coded as soon as he or she walks in the door. The codes are quick and convenient, especially for the purposes of filling out insurance forms and getting reimbursed, but they bear little relation to the complexity of people's lives. Why not, Levine asked, code situations rather than patients—could we have a classificatory code for "going through a difficult divorce"? Of course not, it would be too troublesome. Give the depressed divorcee a "disorder" instead and a fix-it drug regimen. Even worse, according to Levine, is when children are diagnosed with disorders that contain the word pervasive. "It's like a death sentence at the age of two. But of course it gives the doctors and professionals total control over the family from then on."

Essentially, it's an interlocking system. A plethora of newly coined labels is sanctified by the *Diagnostic Manual*. And the Manual, in turn, justifies the careers of tribes of specialists (each one an expert in a single label and each one scrupulously loyal to the Manual), which, in turn, makes the job of the mental health care system more streamlined while also legitimizing a vast consumption of drugs. Everyone is happy, so to speak.

"I see this all the time. You put the child on Ritalin because he's 'difficult.' Then as he gets older the drugs wear off and you declare that he has Obsessive Compulsive Disorder. So you give him different drugs. Then those drugs wear off and you say he's bipolar depressive: a new round of drugs. And so on." But depression is a response in the individual to intricate problems that have to be faced; it's not a disease like kidney failure.

"I think," Levine sighed, "that in this country we always simplify everything to the maximum degree. So we simplify the suffering individual. We make him into a material malfunction."

I had already noticed a tendency among people with syndromes to use the verb "to be" in describing their condition. Instead of saying, "I have Attention Deficit Disorder" they'd say, "I am Attention Deficit Disorder." The disorder becomes the man. Imagine, though, someone declaring, "I am renal failure."

"Perhaps," Levine concluded, "we need to get back to a more humanist way of dealing with people. Just describe the patients as they actually are."

I thought to myself that this might be a good way to proceed in my own journey in pursuit of the enigmas of normality and its opposite. For "Asperger people" are essentially as baffling today as they were to Hans Asperger himself. "The path to understanding," Asperger wrote, "necessarily begins with the individual himself . . . it looks for parallels between an inner region and an outer one."

This is a metaphysical quest, not a biochemical one. It admits that a biological inner shape is always meshed with an outer world, a culture. The afflicted individual is not a bundle of neurological problems. He or she is a story, a kind of tale—a narrative made from the epic conflict of two hostile principles. If this is a truism to any sophisticated psychiatrist, it still needs emphasis in a culture long sold on the putative miracles of pharmaceuticals that bear an eerie resemble to soma, the happiness-inducing drug of Aldous Huxley's *Brave New World*.

CHAPTER I

ASPERGER AND I

This other man is also a human psychologist: and you say he wants nothing for himself, that he is "impersonal." Take a closer look!
—Nietzsche, *Twilight of the Idols*

I had lived in New York for many years before I became aware of the nocturnal adventures of Darius McCollum. My awareness came slowly, passed on in anecdotes and bits of gossip exchanged among friends. It was advanced by the occasional tongue-in-cheek story on TV, but one never saw Darius McCollum's face. "Tonight," one would be told, "Darius McCollum has struck again." But in fact Darius was never seen as he went around the city playing his pranks. He liked to dress up as an employee of the New York City Subway. Carrying a two-way radio, with a fake uniform and ID, McCollum infiltrated stations, construction sites, and trains, passing himself off as a crucial cog in the system. He was a bit like Ronald Biggs, the infamous and much-loved Great Train Robber of the 1960s—a hit-and-run conman, except that in his case the heist was little more than an act of impersonation—no one was ever harmed, and no profit was ever scored. Darius simply loved to pass himself off as a subway official and go about business that clearly was not his own. He was a one-man fifth column, but with nothing to subvert and nothing to steal.

Gradually, Darius McCollum's exploits assumed a dashing underground character. One heard of him from afar, through the grapevine or through items in the newspapers, some of whose writers could not conceal a noticeable irritation with their subject. He was compelled to commit fraud, but not for any reason that could be explained. In 1981, when he was fifteen, Darius was apprehended at the controls of an E train at the World Trade Center, having bluffed his way into the driver's cabin with unusual sang-froid. He mildly explained at the time that he had always wanted to drive a train. After that, his exploits only evolved. Dressed in stolen Transit Authority uniforms, he cheerfully collected fares from travelers, signed out two-way radios from transit service centers, cleared trash from tracks, single-handedly put out subterranean fires, finagled the keys to personal lockers, accumulated goggles and mudguards as well as sheaves of official Transit Authority stationery, and most notoriously made off with agency buses and trains. He was, in short, a man euphorically obsessed with trains. "How can I describe it?" McCollum mused. "I like the scenery. I like the schedules." It was a state of self-induced ecstasy.

After more than a decade of sporadic appearances, he had become a minor cult figure. In 1996, a friend asked me if I had heard that Darius McCollum had pulled another stunt. "He's made," he whispered, "another *foray*." A childish excitement gripped us both. A foray into what? Into the intestines of The System, into the rational grid of our simultaneously hated and beloved transport network? Was he then not a one-man fifth column so much as an antic one-man *Monty Python* squad, or a madcap figure like "the Penguin"? At one point the Transit Authority posted thousands of images of his face throughout the entire system, warning its customers to be on the lookout for this dangerous antisocial element akin to the subterranean dregs depicted in *Batman* or, more pertinently, the average hip-hop video. Needless to say, this rash act of propaganda seemed like overkill and to somehow miss the point, and it only added to the urban legend of Darius McCollum.

Every age breeds its eccentrics, and one of ours is Darius, "The Transit Authority's Biggest Running Sore," as one headline had it. On August 24, 2000, the *New York Times* ran a headline that I cut out and pasted into my journal: "Irresistible Lure of Subways Keeps Landing Impostor in Jail." It was the electromagnetic Darius again. The *Times* reporter, Dean Murphy, offered this lead: "If Darius McCollum were a millionaire, he might be excused as an eccentric. If he were a genius, allowances would be made. But McCollum is neither." Darius had been arrested for the nineteenth time, for pulling an emergency brake, on this occasion during rush hour at 57th Street, then coming to the rescue dressed as a transit supervisor cheerfully named either Manning or Morning. Morning or Manning, McCollum's two transit personas were known to many commuters as friendly faces—the human side of the city's subway system. But such an invention of nonexistent characters was judged inexcusably criminal by a New York court, and the thirty-five-year-old impersonator from Queens was sent to prison, after a sentence passed down the following March with a bracing lack of humor, curiosity, or even ordinary humanity. McCollum did not fit anywhere on the stifling psychiatric grid. Interviewed later on Riker's Island, where he sat on an inmate grievance committee and worked as a volunteer typist, he cried with pitiful force, "I am not insane!"

It's a *cri de cœur* that resonates with many. Convinced that we are not in fact insane, we nonetheless ponder our similarities with Darius McCollum. After all, the delight which his exploits evoke in us could not be possible unless we see something of ourselves in his vagabond shoes. Darius escapes into his two characters Morning and Manning in much the same way that a man might fake his own death and reappear glamorously in Peru under an assumed name. Such charades are irresistibly romantic. Who does not dream some time of slipping away from the dreary and pedestrian persona they have come to inhabit, often for no good reason? Our daily life, like our habitual ego, is usually something which we have fallen into almost by accident or through inertia—to want to escape it is hardly reprehensible. Furthermore, we have no scientific proof that we are not insane, nor do we possess any scientific proof that Darius McCollum *is* insane or *not* a genius. "I'm just the average guy," McCollum remarks disarmingly, "who likes to help out."

But if Darius McCollum is not insane, in what way is he not normal? The answer is a surprise. Darius is not normal because he has a neurological illness called Asperger Syndrome, an obscure high-functioning variant of autism that mainly affects boys. In Dean Murphy's second *New York Times* piece on McCollum, it appeared that Darius's lawyers had decided to enter a plea of insanity in order to spare their client jail time. "McCollum," wrote Murphy, "though skeptical, says insanity is worth a try." And this plea of insanity, the claim that Darius suffered from Asperger Syndrome, seemed at one stroke to offer a satisfying explanation for behavior that was so egregiously bizarre. Not only was Darius McCollum an urban legend, he was also apparently the "urban legend of Asperger's." Children with Asperger Syndrome all over the country followed his antics with a special zest, seeing in his hit-and-run escapades a rebellion against what autistics often call the dreary world of "neurotypicals"—that is, the "normal" world inhabited by the likes of you and me. Darius, by contrast, was one of them, part of a special group, a subpopulation who call themselves "the Mindblind," since they have trouble imagining, or even believing in, other people's minds.

One might assume from the case history of Darius McCollum that neurotypicals (or NTs) and the Mindblind are at war with each other. At the very least, their worldviews conflict. Darius regarded himself as an occasional philanthropist, a helping hand, while the New York City Transit Authority saw him as a menace to public safety. There seems little way of reconciling these two points of view, which only reinforces the feeling among Asperger people that they live on a planet, totally different from our own. What always haunted me about Darius McCollum was whether he enjoyed residing on this alien planet or whether he felt like one of the hapless stranded humans on the Planet of the Apes. There was no way of knowing, because never in a million years would Darius ever be asked that question by a professional psychologist, let alone a judge in the Criminal Division of the New York State Supreme Court.

And meanwhile I ride the subway in New York in perpetual hope of one day hearing the rasp of the emergency brakes. If Darius ever escaped from jail or beat his rap, could we, his silent fans, actually one day catch a glimpse of the slight black man dressed as a solicitous guard peeping a phony whistle, tapping his mudguards and walking with calmly measured steps toward an accident that in reality had never happened?

★

Asperger Syndrome remained almost unknown to the general public until the publication of Oliver Sacks's best-selling study of autism, *An Anthropologist on Mars*, in 1995. Sacks's book was not a study of Asperger Syndrome *per se*, but it did contain a complex and extensive portrait of Temple Grandin, an animal scientist with Asperger's at the University of Colorado. Grandin is in a female minority not just in her affliction with Asperger's, but in her line of work. She designs slaughterhouse facilities for cattle, and it was she, indeed, who once called living among neurotypicals something akin to being "an anthropologist on Mars." For his part, Sacks wrote, "Autism as a subject touches on the deepest questions of ontology, for it involves a radical deviation in the development of brain and mind." There was, he wrote, a core problem, a triad of difficulties, namely an impairment of verbal and nonverbal communication, difficulties with play, and a pronounced lack of normal social interaction. But stealing trains is not part of the equation, any more than is a brilliant aptitude for designing slaughterhouse facilities. Darius McCollum and Temple Grandin may or may not have the same disorder, but both have an obsessional interest, a single dominating passion, and both hover around this triad of impairments.

Sacks warns that our understanding of autism is both patchy and slow to evolve. Moreover, it's impossible to know with certainty whether Asperger Syndrome is a type of autism, a subcategory, or whether (as neurologist Isabelle Rapin suggests) it is a completely separate disorder, biologically distinguishable from autism. Nor do we know how to fit those individuals who simply have autistic or Aspergerish traits, those people traditionally seen as mildly eccentric, into a grid of definitions and discriminations. Sacks suggests that the main difference between Asperger individuals and other autistics is that the former can reflect upon and express the nature of their condition, while the latter cannot. In other words, Asperger people can send back reports from the inner world of autism, while more severely affected autistics remain locked inside themselves.

Grandin herself certainly fits this notion of the eloquent, reflective autistic who can describe her inner isolation, since she has written one autobiography called *Emergence: Labeled Autistic* and another called *Thinking in Pictures: And Other Reports of My Life with Autism*. Both became American bestsellers. On the back of the latter we are told that Grandin has designed one third of the livestock-handling facilities in the United States. What is remarkable about her slaughterhouse designs are their "curved-chute systems"—looped ramps that lead the animals toward the slaughtering station without their being able to sense what is about to happen. Seen in aerial photos, the facilities have a lovely geometrical simplicity—in fact, Grandin calls some of them "my ground sculptures." But they are designed expressly to induce calm and quiet in the animals: ramps, restrainers, and pens are all there to quell the anxiety of the doomed herded beasts. Sacks writes that Grandin's uncanny feeling for bovines is as much a feature of her Asperger's as is her total alienation from people and her complete inability to comprehend concepts such as aesthetic beauty or fictional narratives. In one of her books, we see photos of her sitting calmly among cows; but to Sacks she admits that she cannot make heads or tails of *Romeo and Juliet*. "I never knew what they were up to," she confesses. "Almost all of my social contacts," she has written even more sadly, "are with livestock people."

Driving through an idyllic landscape, Grandin will admit that she cannot understand why people call it "idyllic." She recoils from human touch. When Sacks asks her about boyfriends, she shudders and invites him instead to see the "squeeze machine" she has constructed in her bedroom. Grandin lies in this adapted couch and, using remote controls, applies padded panels to her body that soothingly squeeze her in simulation of a human embrace. The squeeze machine makes sense to her, but humans do not. Her high intelligence, with its extraordinary capacity for attention to detail, cannot extend to sexual intuitions. The mental sufferings of doomed livestock are more accessible to her than the preliminary advances of an amorous man.

What Sacks showed in his study of Grandin was that certain autistic people felt as if they were trying to live on an alien planet whose customs, psychological laws, and ways of feelings were disturbingly incomprehensible. What could one do with these Martians but study them anthropologically? Or steal their trains? Evading their embraces might be time-consuming, but each Asperger person seems to build something equivalent to the squeeze machine.

★

The life of Darius McCollum is rife with Aspergerish twists. He has never held down a steady job. He met his fiancée Nelly Rodriquez, a costume designer from Ecuador, on a subway platform (where else?), but his nineteen detentions make relationships a little trying. His mother, Elizabeth, who now lives in North Carolina, remembers that often Darius did not come home at night. "He would run right down there into the subway. I would go down there and look for him, but the workers would say he wasn't there. But he was there." The transit staff, in fact, seemed to relish him and often made him welcome. Every time her son was a no-show, she would write the word OUT on her calendar. Endless therapy and psychiatry failed: every night after school he would descend like Orpheus into the tunnels. By the tenth grade, he had definitively dropped out. "I'm obsessed with trains," he has said. "That's just it." It's a compulsion that our culture is not equipped to accommodate, unless it takes the form of train-spotting, limited purely to observation. For what we distrust most are *active* obsessionists. These nuisances must use the insanity plea if they are not to pay dearly for their unusual passions.

Collecting these bits and pieces about Darius McCollum's career—in and out of crime, in and out of "mental health"—I had to wonder what sort of story such a life must add up to. Psychiatry sees its patients as case histories, viewed in a series of sequential snapshots. Each evaluation, each meeting with the psychiatrist, forms a snapshot that the doctor can then evaluate with reference to a scientific model, a medical model. One can well imagine the futility of such a method in the case of Darius (it had failed miserably, anyway). I had the sense, looking at these odd clips from newspapers about a person I'd never met, that the so-called mental patient, in this case the Asperger case history, is not a case history at all but a story whose implications are difficult for outside observers to fathom. In a sense, of course, all human lives form such stories, in the widest sense of that term. A person will always look at his or her life as a kind of long, eventful tale with a beginning, a middle, and an end, with a coherence, a kind of independent logic and drama all its own. But this is not a fact that a medical model, a psychiatric model focused on curing mental illness, gives much credence to. There is inevitably a profound contradiction between the way we look at our own lives and the way that science does. Can anyone really understand what role subway trains play in Darius McCollum's inner tale? Can anyone say with certainty that the right thing to do, the *curative* thing to do, is to cut him off from them completely?

If Asperger's is a neurological deficit, as most experts claim, then such people have always existed. And many Asperger activists will duly tell you that in prior ages they were regarded as shamans, holy men or diviners, not simply as oddities. It is only the Age of Science, they go on, which has scrutinized them coolly, puzzled over their eccentricities, and judged them fit for various forms of incarceration and treatment. Even allowing for the usual romanticizing of the past, there may be some truth to this. If, as Neil Postman says, America is the first society to be totally dominated and defined by twentieth-century technology, our normality is almost bound to be ever more ruthless and narrow—the normality, in other words, of engineers. Necessarily, it's a hostile place for the Darius McCollums of this world, who are doomed to sit in places like Riker's Island, protesting like so many minor King Lears, "I am not insane!"

Then again, their putative insanity is not always the point. Darius was not in Riker's Island, however temporarily, or more permanently in a penitentiary far upstate, hundreds of miles away from his beloved trains, because he is insane. He is there, instead, because his obsessions disrupt subway timetables. And this brings up another question. How could Darius's passion for impersonating Transit Authority personnel, carefully imitating their dress and manners, be realistically accommodated by a benevolently technological culture? How could the psychiatric czars of our booming normality appease the inner demons of Darius McCollum, hell-bent as they are on pulling the emergency brake during rush hour and coming gallantly to the rescue of little old ladies? The aforementioned czars have no idea.

★

British writer and autism researcher Uta Frith has written that a high-functioning autistic individual often bears an uncanny resemblance to the mermaid in Hans Christian Andersen's disturbing fairy-tale of the same name. In love with a human prince, the mermaid desires to acquire a human form. But this metamorphosis can only be achieved at a daunting cost. In order to gain legs, she has to give up her voice—an unfair bargain at the best of times, but mermaids have no choice when it comes to exchanging flippers for feet. When she finally acquires these human appendages, the disconsolate mermaid finds that walking around on them is "like walking on knives." To make matters worse, she can no longer explain her inner life to the fickle prince, who promptly abandons her and marries someone else, presumably someone who can speak Danish. The fish-maiden's metamorphosis into a woman, in other words, is painful but it's a remarkable failure all the same. The mermaid ends up where she had begun; indeed she is worse off, since now she can no longer swim (or have romances with more compatible mermen).

The question of normality haunts doctors who deal with subtle and fluid psychological conditions like Asperger's. Since there is no agreed-upon diagnostic for this form of autism, psychiatrists rely on hit-and-miss observations of behavior to diagnose it. Frith writes:

> What, after all, is normality? Given that there is an enormous range of social behavior with many degrees of adaptation and success or failure in the normal population, where does normality end and abnormality begin? Does it make sense to talk about deficits and exclusive categories? Should one instead talk about normal and abnormal shading into each other? To put it another way, should one look at Asperger Syndrome as a normal personality variant?

Frith then gives us a case history known as "The Lonely Cyclist." James Jones, a sixteen-year-old English boy, was an only child. Although as an infant he would lie in his carriage and gurgle happily at trees, James was oddly distant and unaffectionate, and for a long time he spoke a language only his mother could understand. He had a tendency to arrange his toy cars in endlessly long lines. At school, he was disruptive, restless, and uncooperative. And at sixteen, he was something of a loner, socially clumsy and irritating to others. He often stole cricket bats during other boys' games, arousing uncontrollable fury in the players.

Apart from memorizing maps of the London Underground, James especially liked taking long lonely rides on his bike. But it was the bike itself that aroused his parents' psychiatric suspicions. One day, on one of his interminable cycle rides, James rode to a local supermarket, piled a basket high with assorted goods, and then walked blithely out, in full view of the employees, without paying. Apprehended a few minutes later, he mildly explained that he merely "wanted a bit of fun." Showing no guilt or consciousness of having done any wrong, he baffled the bobbies with his winning naiveté. Answering a few subsequent questions, he revealed that his mother was blond and "looked American" and that his sole plan for his own future was to live with a schoolmate whose father owned a traveling fair and was a millionaire. But James exhibited none of the usual tics of autism: no odd bodily movements, no special stiffness of movement or lack of gesture. He was a perfectly normal, friendly English boy with a subtle penchant for theft and a noticeable inability to make friends. He was also good at jigsaw puzzles.

Frith then offers this strange insight:

> There is no reason to suppose that behavior shading into normality will ever cease to be a problem for diagnosis. . . . The categorical distinction of normal and abnormal functioning of mental processes at a deeper level can only be inferred and cannot be directly derived from the initial observation of behavior. As far as the naive observer is concerned, anything can happen.

It's an arresting admission. Some individuals, Frith seems to be saying, have a foot in both lands. But what about us? For that matter, what about *me*?

★

It was reading the case history of the Lonely Cyclist that first gave me an uneasy feeling about some of the tendencies in current psychiatry. Occasionally I had picked up the thick *Diagnostic and Statistical Manual of Mental Disorders*, put out by the American Psychiatric Association and familiarly known as DSM-IV, and browsed in disbelief through the hundreds of newly invented syndromes for things that to me seemed perfectly ordinary. In psychiatry, writes Dr. Peter Szatmari, who has written extensively on the challenges of defining and classifying Asperger's diagnoses, "building a classification system is in many ways like building a house, although too often the house is built on shifting sands."

Should we be suspicious of psychiatric expertise if it strikes us as *immoderate* in some way? It's a difficult question for the humble layman to grapple with, for our ignorance makes us feel guilty for having simple-minded prejudices in this domain. But the feeling of disillusionment is nevertheless widespread. In his hilarious 1988 memoir-novel *Wittgenstein's Nephew*, for example, Thomas Bernhard recalled his friendship with Paul Wittgenstein, the famous Viennese philosopher's obscure but brilliant nephew, an erudite Viennese dandy and opera buff who suffered from compulsions and obsessions that were stubborn and baffling enough to get him landed repeatedly in Vienna's Am Steinhof mental asylum. In one episode the Aspergerish duo of Bernhard and

Wittgenstein set off across Austria in a vain attempt to find a copy of the *Neue Zürcher Zeitung* newspaper in which a particular review of Mozart's *Zaïde* had appeared. Madly driving from town to town in their search for the high-brow paper, they are driven insane by their inability to find a single copy of the *Neue Zürcher Zeitung*, the reading of which has suddenly become to them a matter of life and death, and the absence of which has become a sign of Austria's cultural decay. Reading this, I couldn't help feeling that Paul Wittgenstein and Darius McCollum were actually kindred spirits, splendid imps of the irrational. And the psychiatrists?

> The so-called psychiatric specialists gave my friend's illness first this name and then that, without having the courage to admit that there was no correct name for *this* disease, or indeed for any other, but only incorrect and misleading names; like all other doctors, they made life easy for themselves—and in the end murderously easy—by continually giving incorrect names to diseases.

Then Bernhard goes into one of his magnificent riffs:

> Of all medical practitioners, psychiatrists are the most incompetent, having a closer affinity to the sex killer than to their science. All my life I have dreaded nothing so much as falling into the hands of psychiatrists, beside whom all other doctors, disastrous though they may be, are far less dangerous, for in our present-day society, psychiatrists are a law unto themselves and enjoy total immunity. Psychiatrists are the real demons of our age. . . .

I laugh out loud reading these passages, but not without a certain crawling of the flesh because Bernhard is sniping at a deep unease, which I for one have felt for a long time. In the case of what are called Pervasive Developmental Disorders in children, or PDDs, a vast conceptual quagmire can be clearly seen in the pages of the *Diagnostic Manual* itself. "Rules of evidence," Szatmari goes on, "have proved relatively unsuccessful in establishing clear differences *within* the major subcategories of disorder." It's an understatement of thrilling serenity.

For as it turns out, many of these subcategories seem themselves dubious. Conduct Disorder, Defiance Disorder, and, most famously, ADD, Attention Deficit Disorder (or ADHD, Attention Deficit Hyperactivity Disorder, as it is now bizarrely known) are now being increasingly criticized as pseudo-medicalizations of behavior that nearly all children share. Psychiatry covers its tracks by admitting that there are ubiquitous "overlaps" and "spectrums," but the harsh fact remains that none of these disorders has been shown to be rooted in that same biology which American psychiatry otherwise so ardently worships. "Nevertheless," Szatmari claims, "clinicians feel that, by and large, these clinical distinctions are *useful* without necessarily being valid." It's an utterance worthy of the Sphinx.

The DSM-IV lists five disorders under the umbrella of PDDs. Aside from autism and Asperger's, we find Rett's Disorder, an affliction mainly of girls that leads to "loss of social engagement" as well as an alarming deceleration of head growth between the ages of five and forty-eight months. Loss of hand skills is also one of its foremost symptoms, along with severely impaired "expressive and receptive language development."

Then there is the terrifyingly named Childhood Disintegrative Disorder (CDD), also known as Heller's Disease. For sufferers of this forbidding condition, the symptoms are also primarily social: they include loss of "expressive and receptive language" and bladder control (are the two connected?), inability to play, lack of "social skills," repetitive language and "stereotyped patterns of behavior," and lastly "lack of emotional reciprocity," whatever that is. The clinicians also note a typical inability to form peer relations.

Finally we come to a kind of catchall syndrome intended to include all these children who cannot be labeled according to the above. This is the preposterously named Pervasive Development Disorder Not Otherwise Specified, or PDD-NOS. Notes the Manual: "This category should be used when there is a severe and pervasive impairment in the development of reciprocal social interaction, verbal and nonverbal communication skills . . . but the criteria are not met by a specific Pervasive Developmental Disorder." One can only wonder who sets the definition of "normal nonverbal interaction." Or whether, if a test for it were ever imposed upon most of the population, anyone would actually pass it. The National Institute of Mental Health even names a difficulty in learning math as a mental ailment called Developmental Arithmetic Disorder—a severe illness from which I have suffered all my life. I must also confess that, in addition, I have suffered from what the Institute calls "co-ordination disorders that can lead to poor penmanship."

★

Yet the case of the Lonely Cyclist was different. For above and beyond the byzantine corridors of psychiatry and its submerged paranoia and politics, here was the clinical description of a lonely English boy who could easily have been myself and, in many respects, actually *was* myself. In young James's symptoms I found several things I recognized from my own adolescence. I, too, had been an inveterate solitary bike rider. I, too, had sequestered myself in my room for hours on end, just as James was reprimanded for doing. And I, too, had stolen things from supermarkets just to alleviate boredom, deliberately allowing myself to be caught in order to pretend to be disturbed—it was a game I loved playing with humorless corrections officers and school psychologists. The incessant bike riding didn't seem unduly abnormal in itself, either, although it swallowed up countless days; and I have no memory of stealing cricket bats. Yes, there were own goals (accidentally scoring for the opposition), but they were perfectly explicable from the point of view of psychology, let alone simple physics. What was less explicable was my hatred and terror of other people, especially my parents' friends, not to mention my own.

I remember my ninth birthday party, to which some thirty children were invited, complete with a professional clown and fire-eater. This was in a small town in Sussex called Haywards Heath, a stiflingly closed world where everyone knew everyone. Birthday parties were in reality a means for suburban parents to spy on the social progress of other parents, not to celebrate the anniversaries of their offspring's birth. Exhibiting what I now see might be termed an Aspergerish trait, I locked myself in my room, tied together two sheets and let myself down through my bedroom window onto the roof of the adjoining greenhouse. Unluckily for me, I smashed through a glass pane as I was descending, and my father caught me just as I was scrambling over the garden hedge in a desperate attempt to escape from what was supposed to be my own celebration. This bold breakout was greeted with a particularly venomous disdain. After all the trouble my parents had gone to in order to make me happy, inviting over all my little friends, and I had bolted at the first opportunity and tried to vault over the hedge into Mrs. Burns's back garden. I'd been deeply ungrateful, to put it mildly. Thirty years later, it would be classified in a different light and probably treated with Seratonin-uptake inhibitors. Back then, it occasioned only a mild psychological evaluation and a week deprived of TV.

But it was also noticed that I talked to myself, rode my bike up and down a street called Sunte Avenue dozens of times, always wore dark blue sweaters, was obsessed with *Star Trek*, ran private clubs at school dedicated to the assassination of certain teachers, ate fish-sticks every day, walked miles to obtain a single brand of yogurt that I loved, and painted my bedroom windows black to keep out the sun. The first enthusiasm I can remember having for anything that required skill was the lute. Yes, the lute—not exactly a common hobby among suburban schoolboys. I liked the lute because it was complicated, esoteric, and nobody knew what it was. It had a crazy system of tuning pegs that defied common sense and made a droning, otherworldly sound that suddenly evoked for me the grim realities of the Middle Ages. I thought of it as a soundtrack for the Black Death. Apart from this, it was excitingly unplayable, and the small amount of music that I could play on it was seven hundred years old. This made it odd. Its oddness made it sexy, and its sexy oddness made me oddly sexy—or so I thought.

But the lute occasioned a second psychological evaluation, if only because, in frustration over the instrument's complex arrangement of strings, I one day smashed the lute to pieces on the road outside our house before leaving for school. A passing motorist rang our doorbell at eight in the morning and said to my mother, "I've found some pieces of lute in the road, madam. Are they yours?"

★

Psychiatrists famously say that there is a dash of autism in everyone, without revealing how much or why; in which case, I am at liberty to observe such a dash in myself—or anyone else I feel like. In the most general sense autism, from the Greek *autos*, or self, is only a state of separation from others, a self-sufficiency that implies a radical isolation from the usual litany of social engagements such as love, affection, duty, and expectation, but also from fear, guilt, and recrimination. It may be not quite social, as Robert Frost had it, but is, as they say, in the range of normal.

A withdrawal into the self seems actually to be a habitual human retraction, and its reasons are complex. The romantic celebration of solitude lies dormant even in the harried citizens of the technological state. And in them the craving for isolation will take the form of obsessive compulsions, personality disorders, or simply a lust for long pointless bike rides. In my case, it also took the form of exactly those kinds of ridiculous phobias that are today taken as a sign of Asperger's, or worse: a fear of things like birds, telegraph poles, or snails. When I first saw Alfred Hitchcock's *The Birds* I went into a near cataleptic fit: it seemed to be the incarnation of every phobia I was capable of. It is strange, too, to note how the humans in this film masterpiece are trapped inside their useless technology—a car, a telephone booth—by a primal force that they cannot control: the birds. They are forced by these avenging crows and gulls into being involuntary autistics, and the terror of the film lies in this claustrophobia. In the scene where the character played by Tippi Hedren is trapped in a telephone booth as swarms of gulls batter the glass panels around her, a moody silence falls in the soundtrack. The human victim trapped in her booth becomes blinded by a confusing maelstrom of bird noises and by a sickening crunch of glass: the world as an autistic person might feel it, blinded or deafened by an onslaught of violent sensations.

One of the significant features of Asperger's is what is called "sensory overload," or an inability to cope with diverse sensations or information pouring in through the senses. Acutely sensitive to color, sound, and motion, Asperger people suffer a kind of kinesthetic breakdown at certain moments of charged sensory intensity. The term "nightmare" for them usually implies a crisis of the senses, a shutting down of their nervous system in the face of excessively vivid sensations. Few of us ever experience this breakdown. But if the world suddenly became abnormal and we were attacked by birds, we might. Like Asperger people, we might suddenly notice the repulsiveness of certain fabrics or even certain liquids, metals, or fruits; we might recoil from overdone sensations, from neon logos whizzing around our heads or the sound of subway car brakes. It's easy to see why autism could often have been taken for saintliness in the past: it's nothing less than a harried retreat from the overwhelming world of the senses.

But in the case of Asperger's it is also a mechanical orderliness, a mania for repetitive actions. These compulsions are often simply exaggerations of manias that the rest of the population suffers from. Throughout my childhood, for example, I had a fixation with the cracks between sidewalk paving stones. In England, sidewalks are neat and orderly and the cracks occur at nearly exact intervals. My fixation took the form of believing that if my foot landed on one of these cracks a curse would seize me, as if the paving stones themselves would take note of this infraction of the crack rule and hold it against me. I would always take care, therefore, to step nimbly over the cracks, even if it meant contorting my gait accordingly, shortening or lengthening steps unnaturally. Over time, this fixation became so deeply ingrained that I could navigate around paving cracks almost unconsciously, as if a sixth sense were guiding me from above and saving me from certain catastrophe. What this catastrophe might be I never paused to ponder. What could a series of paving stones do to me? Vaguely, I thought that it might simply be a kind of animist displeasure.

Another fixation of a similar nature: Before the age of ten I would say goodnight to every object in my room, beginning with the lamp shade above and ending with individual nails in the floorboards below. Over time, this goodnight taking became so fanatical that it encompassed motes of dust on the windowsill and the tube of toothpaste sitting on the edge of the basin. What mattered to me was not to leave anything out, because every object had its tender feelings. If I left even a toothpaste tube out, it would be wounded, hurt, or humiliated, and bad things would come of this accidental exclusion.

It was the same with lampposts. If I walked around a lamppost counterclockwise, I'd have to go back immediately and "unwind" myself from it. It was as if I'd draped a piece of my own intestines around the pole which tugged me back toward it like a tough elastic band: impossible to keep on walking and ignore the pull back toward the lamppost. I'd have to disentangle myself from this imaginary rope that tied me to the lamppost, because I had walked around it counterclockwise instead of clockwise. A troubling scene, therefore, greeted my mother whenever we went out walking. Her son could be seen walking counterclockwise around every lamppost that he passed, then retracing his steps and going clockwise around the same lamppost. I've since seen in clinical literature that many children have these fixations—reading human qualities into objects—and that they are now usually regarded as signs of Obsessive Compulsive Disorder, or worse.

In the 1960s, however, these tics were not pathologized. Children were allowed to have tics in the first place, and the incomprehensible meanderings of their imaginations were left intact. This was because childhood itself was still intact, much like the symbolic system of a remote Papuan tribe surviving into the twentieth century. Since then, it seems to me, the Papuan tribe has surrendered its unique cosmology to anthropologists, while childhood has surrendered its unique inner world to psychiatry. Both have more or less disappeared.

★

If we cast a suspicious eye back on childhood eccentricities that once seemed merely innocent, how much more severe do we have to be with our adult selves? "The child is father of the man" now seems a sinister adage. The adult person cannot wriggle out of fixations so easily because he or she structures his or her waking thoughts too much. I've often wondered if the only difference between myself and Darius McCollum is that Darius has the shamelessness to actually act out his fixations rather than just think about them. It's not that I could ever be fixated on trains, and my lute phase is definitely well over. It's just that I'm disappointed to find that the fixations I do act out are somewhat trivial.

When I'm in strange hotel rooms, for example, I'll feel a gnawing craving to turn on the TV and watch the *Iron Chef* contests from Japan. In these staged culinary battles, Japanese master chefs specialized in different cuisines (French, Chinese, Japanese, etc.) and known as Iron Chefs are pitted both against each other and against challengers in trying to concoct outlandish meals around a single theme—say, octopus or pitted prunes. They work feverishly against the clock and against a running commentary robotically translated from the Japanese. The whole effect is infantilely gripping, not to say vilely mouth-watering, and I've found that *Iron Chef* is also a great hit among Asperger people, who relish the sudden creation of squid ink ice creams and asparagus fondues out of nowhere. For me, however, the *Iron Chef* theme is not so much a fixation as a kind of addictive calmant. But if I stay in that hotel room for long enough there is a practically one-hundred-percent certainty that by midnight I will be watching Iron Chefs dueling over their extemporized gothic menus.

I can honestly say that watching *Iron Chef* has taken over completely from paving cracks, lutes, and lampposts, and that knowing with exuberant pedantry every detail of these ludicrous shows is a psychologically binding rite for me—from the colored satin costumes of every Iron Chef to the impassive faces of the various moderators and the history of each legendary duel, pitting Chef Sakei against Chef Miura, the Italian maestro casually known as the Master of Garlic—who won the battle of the peaches, who lost the famous war of the garlic, who scored a knockout victory with the squid ink ice cream.

Over the years, I've built up an entire internal pedantry around this show and am easily able to rehearse in memory every menu I've ever seen performed on it. On planes or in airports, I frequently draw up in my mind one of Sakei's legendary Rabelaisian menus and proceed to mentally eat it over and over again while replaying the caustic accompanying comments of gorgeous TV actress Fumie Hosokawa or the solemn food critic Dr. Yukio Hatori. All of this, I suppose, makes me rather Aspergerish.

The same could be said of airports. Whereas most people get in and out of them as quickly as they can, I love to linger in them as long as I can, savoring their geometries, cool air, and glassy surfaces. Among all the Asperger people I have met, I have found several who are also drawn to these qualities of airports and chosen TV shows. I have met seriously dysfunctional young men who can recite entire episodes of a 1967 episode of *Peyton Place* or six consecutive ones of *Frazier*. True, I have never encountered a fellow aficionado of *Iron Chef* with quite my level of attachment, but lovers of airports are as common among Asperger folk as are a loopy fascination with traffic systems or outmoded computer programs. Inexplicably to ourselves, we are drawn to things that resemble clockwork, that have a kind of inner logic which defies common sense.

All of which leads me to conclude that, although I think I know I have never had Asperger Syndrome, I am not at all sure that I do not have some Aspergerish traits. I may, in fact, have far more than I'd care to admit, living a life that is every bit as Aspergerish as it is qualitatively normal. And it's this very possibility that drew me to the obscure figure of Hans Asperger in the first place. For it was this unassuming Viennese doctor who first suggested that we are all a little autistic, a little Aspergerish, and that the line separating us neurotypicals from *them* is little more than the kind that is drawn in sand with sticks.

★

Not much is known about Hans Asperger (1906–1980). He joined the staff of the Vienna University Pediatric Clinic as a young man, specializing in the novel field of *Heilpädagogik* (therapeutic pedagogy). In 1935, he became the director if the clinic's *Heilpädagogik* department. Photographs from the period show a frank, earnest face, glamorous in its way, with a full head of curly hair and intense spectacles. He was immediately drawn to children who were misfits, outsiders. These were difficult loners who could not get on with people around them. Withdrawn, silent, dreamy, stubbornly lost in their own inner world, they were often described as vicious, socially inept, imbecilic, and obstinate. In the Nazi health system, autistic children were never far from a prospective death sentence, and Hans Asperger may well have been aware that there were grave reasons for rescuing such children from a fateful diagnosis that would label them as "parasites." Did some autistic children, he wondered, have special gifts that made them salvageable? Could an inspired and humane pedagogy rescue them from social oblivion or worse?

Heilpädagogik was one of the most remarkable innovations of early twentieth-century medicine in the German-speaking world—a fusion of empirical rigor, spiritual insight, and hands-on inspiration, which demanded that a doctor working with disturbed children be simultaneously a scientific clinician and intuitive healer. It's not a model that has been eagerly imitated since. At the Vienna Clinic, children were housed in the beautiful Widerhofer Pavilion where, as Uta Frith tells us, the rooms were decorated with friezes and the furniture was designed by architects. Hans Asperger's assistant and inspiration was a nurse named Viktorine Zak, who developed a regime involving music, exercise, play, and speech therapy—a remarkably advanced concept for its day. Rather than test children inflexibly and force them to obey strict controls, the emphasis was on a kind of remedial intuition that was exercised with great gentleness. It is astonishing that such an enlightened hospital could have been running in the heart of the Nazi state.

Heilpädagogik also demanded that doctors be more than technicians diagnosing and managing pathologies. Hans Asperger himself, like Erwin Lazar, his predecessor at the Vienna University Pediatric Clinic, seemed well suited to filling such a quietly exalted role. They were complete humanists, as interested in art and culture as they were in science. The notion that a doctor, or any scientist for that matter, should also be a man of culture at a high level, permitting these domains to filter into each other, now strikes us as quaint, if not downright inappropriate, even bizarre. Nowadays, would we even have faith in any expert who was not narrowly and obsessively specialized? We demand that our doctors be technicians. But to someone like Asperger, the pretension of claiming to heal minds would have seemed bogus in a doctor not familiar with manifestations of human nature over a wide spectrum. As it happens, Hans Asperger and his colleagues expressed the values of a typically Viennese pre–World War II–culture: skeptical and humane, logical and accessible, wary of obscurantist expertise and fashionable sentimentalism. Their only un-Viennese characteristic was their almost scornful rejection of psychoanalysis. From almost the beginning Asperger had been convinced that the roots of autism were biological, and therefore, he surmised, therapy would be useless. Significantly, the clinic itself was located within the department of pediatric medicine, not psychiatry.

Asperger published his doctoral paper "Die Autistischen Psychopathen im Kindesalter" ("Autistic Psychopathy in Childhood") in 1944. It came one year after the German expatriate Leo Kanner at Johns Hopkins had published his classic paper on autism, "Autistic Disturbances of Affective Contact" in the oddly named journal *Nervous Child* (1943). These two papers came to be seen as the twin pillars of modern ideas of autism, though they mapped out somewhat different aspects of the malady. While Kanner set out to create a definitive description of what is now called "classic autism"—that is, its more severe or mainstream symptoms—Asperger concentrated on the higher-functioning variety. Kanner's classic autism encompassed children who were severely handicapped. Unable to sustain normal social manners, let alone normal daily functions, these children were tragically locked inside themselves. They rocked incessantly in their chairs, knocked their heads against walls, were unable to converse or play. While Asperger's children were often rather normal seeming, Kanner's were radically impaired. Nevertheless, Kanner's and Asperger's two papers overlapped because the groups of children studied by both men exhibited similar autistic problems: a lack of social integration, stereotypical behavior, isolated interests, and poor communication skills.

In a 1956 paper, Kanner gives five diagnostic criteria for autism, as observed in his eleven child case histories. In his own words, they are:

- A profound lack of affective contact with other people.
- An anxiously obsessive desire for the preservation of sameness.
- A fascination for objects which are handled with skill in fine motor movements.
- Mutism, or a kind of language which does not seem intended to serve interpersonal communication.
- The retention of an intelligent and pensive physiognomy and good cognitive potentional manifested, in those who can speak, by feats of memory.

Kanner's children also showed signs of echolalia—a mechanical repeating of phrases or words associated with experiences even as the contexts of those experiences change. For example, a child will suffer a splinter in her finger and thereafter will repeat "Got a splinter" every time she suffers physical pain, whether that pain is caused by a splinter or not. Kanner's children, in fact, suffered many difficulties with language while Asperger's tended to be precociously verbal—indeed, they tended to talk like pedantic miniature adults. In addition, the latter were strikingly expressionless, with a stiff lack of humor. Kanner, meanwhile, thought that autistic children had a disturbed relation to people but not to objects (he thought they were often highly dexterous), while Asperger considered their relations to both people and objects equally disturbed. Frith sums up the differences in the two doctors' perspectives:

> Children who do not talk or who parrot speech and use strange idiosyncratic phrases, who line up toys in long rows, who are oblivious to other people, who remember meaningless facts—these will rightly conjure up Leo Kanner's memory. Children and adults who are socially inept but often socially interested, who are articulate yet strangely ineloquent, who are gauche and impractical, who are specialists in unusual fields—these will always evoke Hans Asperger's name.

Both men repudiated any connection between autism and schizo-phrenia, despite some nagging similarities that would seem to suggest otherwise. (In a 1979 paper, the child psychiatrist Sula Wolff later described what she called "schizoid personality in childhood," observ-ing such things as social isolation, emotional detachment, and an inability to communicate.) As for the popularly understood differences between Asperger's and Kanner's early accounts of autism, it is gener-ally felt that Kanner presents the more pessimistic diagnosis, seeing autism as a disaster for the child, while Asperger looked more on the bright side, seeing in his afflicted patients the glimmers of an unusual, elusive genius.

In his 1944 paper, Hans Asperger gives us the case histories of four boys who have since become the archetypes of the Asperger child. He calls them Fritz V., Harro L., Ernst K., and Hellmuth L. It's clear that Fritz and Harro are the most "Aspergerish" of the four, which is to say the most gifted, and it is in them that Asperger himself shows the most interest.

Fritz V. was a scion of an illustrious family that had produced many of Austria's greatest poets. But at the end of his first day of school his teachers had nevertheless pronounced him "uneducable" and referred him to the Remedial Department of the Pediatric Clinic. Asperger agreed that he had a "pronounced destructive urge," as well as being restless, fidgety, and disobedient. "His eye gaze," he wrote, "was strik-ingly odd. It was generally directed into the void, but was occasionally interrupted by a momentary malignant glimmer." He had a tendency to jump, hit, and echo speech, and could not learn the polite form of address in German, *Sie*. Instead, he called everyone *Du*. Most striking, however, was the fact the Fritz's mother was very similar to her son, which is to say remote, a little robotic, and often disheveled. "In the way she moved and spoke," Asperger noted, "indeed in her whole demeanor, she seemed strange and rather a loner." Like him, she seemed to show a complete lack of intuitive social understandings in her dealings with the world. (The father, on the other hand, was merely pedantic and correct, in a reticent sort of way.) When mother and son walked along the street together, they gave the impression of having nothing to do with each other.

Asperger had clearly already decided that autism, and by extension Asperger Syndrome, runs in families, a notion that is now widely accepted. And as proof of this genetic connection he mentioned the fact that, whenever things got rough at home, Fritz's mother would simply take off for the hills—literally, in her case, for she loved to disappear into her beloved mountains and stay there until her domestic affairs quieted down. In other words, in moments of great stress she herself would behave like a person touched with autism. Kanner, too, had noted similar alienations between mother and son in his case histories and had raised the question of "the autistic family." All of Asperger's four boys seemed to have come from eccentric, isolated families where individuals existed in their own solitary orbits, hardly interacting at all. The father of Ernst, for example, is described as "clearly eccentric and a loner." These families must have been a bit like that of the theoretical physicist Stephen Hawking, all reading their own books around the evening dinner table in total silence. Such families would be difficult to grow up in, and their sons might well inherit baffling abnormalities.

The four were certainly offbeat. Asperger noticed that all of them had a faraway look in their eyes, a peculiar lack of focus. Like Fritz, Harro had a gaze that was "far away," while Asperger says of Ernst, "The eye did not seem to grasp anything and was vaguely aimed into the distance." Time and again, Asperger says that these boys looked as if they "had fallen from the sky." Ernst, in addition, possessed a kind of effeminate ornateness. "His voice, too, fitted in with this. It was high, slightly nasal and drawn out, roughly like a caricature of a degenerate aristocrat, for example the immortal Graf Bobby." The latter was a comic character popular in the German-speaking world between the wars, an absurd upper-class nincompoop completely bewildered by ordinary events.

Of the four, the strangest physically was Hellmuth. Asperger gives this haunting description of him:

> His appearance was grotesque. On top of the massive body, over the big face with flabby cheeks, was a tiny skull. One could almost consider him microcephalic. His little eyes were set close together. His glance was lost and absent but occasionally lit up by malice.
>
> As is to be expected from his whole appearance, he was clumsy to an extraordinary degree. He stood there in the midst of a group of playing children like a frozen giant. He could not possibly catch a ball, however easy one tried to make it for him. His movements

when catching and throwing gave him an extremely comical appearance. The immobile dignity of the face which accompanied this spectacle made the whole even more ridiculous. He was said to have been clumsy in all practical matters from infancy, and has remained so since.

What made Hellmuth even more unnerving was his grave love of poetry. Not only did he love verse, but he spoke all the time as if in meter, using unusual words in equally unusual syntactic combinations, "seemingly full of insight and superiority." This made other children taunt him as they pursued him through the streets. "He clearly," Asperger writes, "did not have any feeling for the fact that he did not really fit into this world."

In reading these case histories drawn from children of Nazi Austria, which have provided the basis for the syndrome that now bears Asperger's name, I am constantly intrigued by the question as to whether they can possibly bear any resemblance to American children of the twenty-first century. Some of Asperger's observations have certainly been artfully brushed under the carpet by American psychiatrists. He reports, for example, a "calculated malice" in every case. Asperger even goes so far as to regard maliciousness itself as a symptom of the syndrome—each boy has that telltale malignant glint in his otherwise unfocused eye. Moreover, he reports that their behavior was erratic, spiteful, and violent. Harro liked to leave his desk during lessons and crawl around on all fours while recounting "long, fantastic stories." He also liked to go into fits of berserk violence or launch into "homosexual acts" with other boys (Asperger actually calls them "coitus attempts"). Fritz himself also had some distinctly unpleasant habits. He would eat enormous quantities of pencils, pieces of wood, and even the occasional fragment of lead. He also had a penchant for licking tables, as well as for jumping into puddles. "He had no mastery," Asperger muses, "over his body." A little later he adds: "To put it bluntly, these individuals are intelligent automata." They were, he says, afflicted by a "primitive spitefulness."

These clinical descriptions have been quietly dropped from homegrown American diagnostics. They are not nice enough and do not conform to what Americans want their children to be like, even if they are admittedly brushed by autism. Licking tables and jumping onto unsuspecting boys with homosexual intent? Even eating fragments of pencil lead would be seen as far worse than being a mere autistic. Instead, contemporary psychologists concentrate on the other side of the classic Asperger child as found in these same pages: intelligence, gifts, aptitudes. The modern view likewise seems to stress that since Asperger children are sticklers for rules of any kind, they tend if anything to be excessively law-abiding. But the original diagnoses tell a story that's not uniformly bright and upbeat.

America's euphemizing tendency is an expression of both a willed optimism and a craving for harmless consensus. There seems almost to be a superstition about naming unpleasant things for what they actually are, particularly if the things in question are ugly, frightening, divisive, painful, or irremediable. Our spooky assertions of artificial optimism and artificial consensus are repeated merely in order to insist that our natures are sunny and that no deep effort of the imagination is needed to understand, and perhaps repair, the world. But behind the bland language of moral propaganda, a dark unease lurks—the fear that the world, among other things, is organically unequal and rent with conflict, and that significant parts of it are impervious to the blandishments of American optimism, know-how, and stick-to-itiveness.

It was in a rather different spirit that Asperger made great claims for his children, often describing them as precocious art connoisseurs or naturally gifted mathematicians. He was not exaggerating or trying to make them feel better about themselves; he had simply observed that if Fritz had no mastery over his own body, he certainly did have mastery over his fractions. At the age of six, Fritz could calculate two-thirds of the number 120 almost instantaneously, as well as grasp the concept of negative numbers, which he seemed to have worked out independently for himself. What was remarkable about Fritz was that his thought, his knowledge, seemed to proceed from the inside out and not the other way around. Fritz, like the three other boys, performed best on intelligence tests when he could react spontaneously to questions rather than when he had to reproduce what he had learned mechanically from others. His mind had a warp of its own. Writing of Ernst, Asperger noted that "his knowledge of the world arose mainly from his own experience and did not come from learning from others."

Harro, too, was precocious at math, but his way of doing it was completely autonomous. At age eight he made mental calculations at extraordinary speed. To explain the logic of the proposition that 27 plus

12 equals 39, Harro said: "2 times 12 equals 24, 3 times 12 equals 36, I remember the 3 (i. e., 27 is 3 more than 2 times 12) and I carry on."

For 34 minus 12 equals 22, he took a similarly circuitous route: "34 plus 2 equals 36, minus 12 equals 24, minus 2 equals 22. This way I worked it out more quickly than anyone else." And it was true. Although his method was far more complicated than necessary, Harro could do calculations far faster than his classmates using the traditional methods of subtraction or addition. Listening to his explanations, Asperger began to wonder if there was a unique "autistic intelligence," or "autistic originality"—a profundity of thought and experience that for some reason was incapable of imitating thought and experience coming from the outside, which is to say from adults. Harro had never been able to learn mathematics from an adult, so he had simply worked it out internally for himself. "Clearly," an impressed Asperger coolly concluded, "it is possible to consider such individuals both as child prodigies and imbeciles with ample justification."

But what characterized this unique "autistic intelligence"? The most interesting part of the intelligence tests that Asperger set for his young charges was what are called comparison exercises. The child is asked to explain the differences between two simple elements such as stove/oven, stairs/ladder, or fly/butterfly. The replies of Harro to these questions particularly intrigued Asperger because they mixed up objective observations with personal memories and feelings in a seamless, dreamlike way. Harro's answer to the fly/butterfly contrast is well worth quoting in full, because it shows what Asperger means when he talks about his children differing in their perceptions from normal children. Harro, he found, regarded objective differences between flies and butterflies on the same level of importance as his own memories of flies and butterflies. As Asperger noted, it was extremely difficult when doing these tests to get Harro interested in anything, but once he became interested in a contrast he wouldn't stop talking about it until Asperger himself brought the interview to an end. Here is Harro's peroration on insects:

Fly/Butterfly

The butterfly is colorful, the fly is black. The butterfly has big wings so that two flies could go underneath one wing. But the fly is much more skillful and can walk up the slippery glass and can walk up the wall. And it has a completely different development! [Harro now becomes very excited.] The fly mother lays many, many eggs in a gap in the floorboards and then a few days later the maggots crawl out. I have read this once in a book, where the floor talks—I could die laughing when I think of it—what is looking out of this little tub? A giant head with a tiny body and a trunk like an elephant? And then a few days later they cocoon themselves in and then suddenly there are some dear little flies crawling out. And then the microscope explains how the fly can walk up the wall: just yesterday I saw it has teeny weeny claws on the feet and at the ends tiny little hooks; when it feels that it slips, then it hooks itself up with the hooks. And the butterfly does not grow up in the room as the fly does. I have not read anything about that and I know nothing about it.

In answer to the same comparison, Fritz replied, "Because the butterfly is snowed, snowed with snow," while Ernest chimed in with, "The butterfly has wings like glass." On another comparison test, that of ladder/stairs, another autistic boy came up with the phrase, "The ladder goes up pointedly and the stairs go up snakedly." (The German is even more lithely expressive: "*Die Leiter geht so spitz und die Stiege so schlangeringelich.*") "It is worth mentioning here," Asperger theorizes, "that all young children have a spontaneous way with words and can produce novel but particularly apt expressions. This is what makes for the charm of child language. Beyond the toddler age, in our experience at least, such spontaneously formed expressions are found only in autistic children." In other words, autistic intelligence finds its expression essentially in a gift for figurative language.

Asperger's charges often came up with surrealist metaphors like:

- "My sleep today was long but thin."
- "I don't like the blinding sun, nor the dark, but best I like the mottled dark."
- "I can't do this orally, only headily."
- "To an art-eye these pictures may be nice but I don't like them."

Their pronouncements are not unlike the effusions of André Breton's automatic writing so popular in the seances of the surrealists—condensed metaphoric effusions of the Unconscious. "As always," writes Asperger, "the miraculous automaticity of vegetative life is at its best when left unconscious."

This vegetative life, as he put it, may have had its miraculous side. But it also produced a darkly enigmatic deviance. Asperger's boys did not have what psychologists call a "theory of mind"—that is, they could not imagine the existence of minds other than their own. The feelings, mental reactions, insinuations, and expressions of other minds could never strike them as real. The reasons for this aberration are unknown. Asperger himself, as we shall see, surmised that the autistic child cannot read the subliminal and preintellectual messages of its mother's face. Compliance, reciprocity, the notion of an Other, Asperger argued, all come from an outpouring of intuitively understood expressions beamed into the child from the mother. No one can scientifically unravel how these infinitely subtle, almost animal, expressions between mother and child actually work. But if they break down in any way, if the flow of emotion and signal is in any way deranged, the child will take off on a tangential development.

★

With the end of the World War II, Asperger's unusual and penetrating work was lost from sight as the center of gravity in child psychology shifted from the German-speaking to the English-speaking world. It was rediscovered only in the early 1980s by a British researcher named Lorna Wing at London's Institute of Psychiatry. Basing her work on a study of autistic children in the London borough of Camberwell, Wing published her findings in the *Psychological Medicine* magazine in 1981. Along with her collaborator Judith Gould of the Social Psychiatry Unit, Wing had discovered that autistic children were indeed highly varied, as Hans Asperger had suggested: some had language and some didn't, some were social and some weren't, and so on. There was not one autism, but many. "We developed the idea," she told me, "that there was a very wide spectrum of autistic disorders of which Kanner's and Asperger's were only a part." In any case, the term "Asperger Syndrome" (AS) was launched in official psychiatry. "I felt," Wing added, "like Pandora opening the box."

Pandora is one of most accessible characters of Greek mythology because her sneaky curiosity is all too pathological. Once I had followed the fortunes of Darius McCollum for a while and then read the paper of Hans Asperger, my own was roused to the same level. In

Camberwell, Wing had found that between 0.6 and 1.1 per 10,000 children had some form of mild mental retardation, but she knew that the actual number of Asperger children was much higher since most of them do not suffer from full-blown retardation. But no one knows how many of them there are. Scientists at Yale's Child Study Unit, including Fred Volkmar and Ami Klin, are uncertain. They observe that the lack of generally accepted diagnostic guidelines for AS means that it is well-nigh impossible to provide more than an estimate of its prevalence. They add that the condition is very frequently misdiagnosed and, if a strict definition of AS is adopted, it is far less common than autism.

Asperger groups talk of as many as 1 in 250 people—a huge exaggeration, but certainly reflective of an exponential increase in the number of lonely, difficult, eccentric people claiming to be afflicted. I began to see it all around me as well—and inevitably, in myself. It is thus with all modern syndromes. Edward Hallowell, author of a popular 1994 book about Attention Deficit Disorder (ADD), *Driven to Distraction*, writes gleefully that "Once you catch on to what this syndrome is all about, you'll see it everywhere. . . . You may even recognize some of the symptoms in your own behavior." A syndrome is conceived (usually in Europe long ago), rediscovered, admitted into the establishment via the *Diagnostic Manual*, and is then let loose into the general population, where it takes on ever more varied, ingenious, and bizarre forms. As the whole notion of individual eccentricity declines in Western culture, we come to rely more and more on the notion of medical disorder and an array of syndromes that can be applied to all who are strange, or simply solitary.

So it was that, walking one hot summer day down a street in Boston, I came across a small theater playing *32 Short Films About Glenn Gould*. Glenn Gould, I thought sullenly: didn't he have Asperger Syndrome? I went in, grateful for the air-conditioning. There was almost no one there; one other viewer seemed to express disgruntled surprise at the sight not of a voluptuous Hollywood diva named Glenn being exposed to cruel indecencies but of a lanky young man trudging across fields of Canadian snow muttering to himself about Bach. I settled in. There was infant Gould sitting on a jetty on Lake Simcoe near Toronto, muttering multiplication tables to himself. Aha, I thought: math, there was a key! A chambermaid at the Vierjahreszeiten Hotel in Hamburg was being interviewed about Herr Gould. Yes, she seemed to be saying, the Maestro certainly was rather strange. Downright bizarre, even. He ate little other than arrowroot biscuits, wore cloth caps and smothering scarves all year round, and wouldn't shake people's hands. Instead, he politely handed them a card that read as follows:

> Your cooperation will be appreciated. A pianist's hands are sometimes injured in ways which cannot be predicted. Needless to say, this could be quite serious. Therefore I will very much appreciate it if hand-shaking can be avoided. Rest assured that there is no intent to be discourteous—the aim is simply to prevent any possibility of injury.
> Thank you.
> Glenn Gould

At his cottage on Lake Simcoe, Gould drove his motorboat around in circles to alienate fish from fishermen. This betrayed a triple fascination—with animal welfare, spinning wheels, and circles. Another key! I began to be convinced. What incomparable, austere solitude! A man alone in a desert of packed ice, trudging under a midnight sun of the North, a Hamlet in a long black coat—the artist as autist.

I wondered to myself aimlessly if perhaps Hamlet didn't have Asperger Syndrome as well. It would certainly explain a lot. The droning soliloquies, the lack of social skills, the obsessive compulsions, the roller coaster moods in the days before Paxil. At the very least one could claim General Anxiety Disorder. One could, I mused, do a whole analysis of Shakespeare diagnosing his characters anew. King Lear? A clear-cut case. Puck? Possibly. Richard III? Definitely. Macbeth? ADD, at least. But whatever Hamlet had, Gould seemed increasingly to me to be a solid case of AS. It seemed distinctly uncontroversial in the light of his manifestly abnormal behavior. I came out of the theater exhilarated. If mental illness was a Shakespearean tale of the wandering individual, the ill-starred life of Glenn Gould was a perfect illustration of its uncanny marriage to the possibilities of genius. Perhaps Hans Asperger had hit on more than he realized.

★

One night I dropped in on one of the monthly Manhattan Asperger parents' meetings, which are held in the Synergia Building on West 65th Street. Although most of the people here are frustrated parents of children who cannot find affordable special education, there would also be—I had been told—a smattering of Asperger adults who attend in order to show solidarity or to introduce themselves to one another. I was not sure, however, quite what to expect. After all, I hadn't yet to my knowledge met face to face with an Asperger's individual. Would I encounter a Rain Man or someone out of an Oliver Sacks case history? Neither was an especially reassuring prospect. In the elevator I wondered if the Mindblind would be as unnerving as I had imagined. I had with me a page I had photocopied from the *American Journal of Psychiatry* to prime myself. It had been written by a fifteen-year-old Asperger's boy and had given me an oddball nightmare the night before:

> My name is Robert Edwards. I am an intelligent, unsociable, but
> adaptable person. I would like to dispel any untrue rumors about me.
> I am not edible. I cannot fly. I cannot use telekinesis. My brain is not
> large enough to destroy the entire world when unfolded. I did not
> teach my long-haired guinea pig Chronos to eat everything in sight
> (that is the nature of the long-haired guinea pig).

The conference room was filled with what looked like affluently white middle-class parents, mostly women, who jauntily launched into a violent jeremiad against the New York City Board of Education. Their tempers were clearly well-frayed through long and obviously Sisyphean struggles against city bureaucrats in what, from the sound of it, must have been one of the most corrupt and inept city agencies in the United States. "This city," one woman lamented, "cannot even educate neurotypical children. What chance do ours have?" This thundering question went unanswered.

After this venting of heartfelt frustrations, the meeting broke up into smaller study groups, and I wandered into a neighboring room. At the end of one of the tables where the parents were hunched, now seized by gusts of laughter, I noticed two individuals sitting a little apart from everyone else. Something about them was immediately noticeable. I couldn't tell if it was the posture of their bodies, the disposition of their hands lying flat on the table in front of them, the cocked angle of their heads, or the shifty eyes that were turned away from the other people in the room. At once, I remembered Hans Asperger's descriptions of the "faraway gaze." I moved a little closer, searching for the concomitant "glimmers of calculated malice." But when I sat next to them, they swiveled their heads like cockatoos and uttered rather weary, amicable greetings.

"I suppose," one of them drawled, "you're from the newspaper?"

He was a bird-like man of around thirty-five, awkwardly slung forward in his chair, and he peered at me sideways, with his head half turned away—precisely the posture that made me think of a bird, as if his vision was not quite binocular.

"Of course he's from the newspaper," the other one drawled back. "Doesn't he look like he's from the newspaper?"

"I don't know, Dennis. What does a man from the newspaper look like?"

They were both catty, effeminate, oddly aristocratic—Graf Bobbys? Their eyelashes, in particular, fluttered constantly, giving them a subtle vibration, like two suspended hummingbirds waiting to dip into some enormous flower.

I admitted that I was from the newspaper. The bird-like man was named Mark Ramoser, and he was a Yale math graduate. The other, Dennis, ran a gay and lesbian radio show.

"So you're here to see the Rain Men," Mark said dryly. His head twitched and he blinked. "Is it what you expected?"

They cackled together and looked bleakly around the room.

"Do we seem normal?" Dennis said.

"Hope you don't mind us asking," Mark put in. "But it's best to get it out of the way first."

I cleared my throat and wondered to myself if the neurotypicals were listening in. "You seem normal," I said, and my voice came out in a high-pitched whine. "Remarkably normal. I mean, perfectly normal. Normally normal. Not inordinately normal. Just normal."

"*Mahulo nui loa*," Mark said.

"Excuse me?"

He leaned forward, very intent, his head still cocked sideways and his eyes averted.

"Hawaiian. I have to tell you, I hate living on the *mainland*."

"The mainland?"

"Yes. Here." He made a vast and operatic gesture around himself.

I said I didn't care for it too much myself.

"The mainland," Dennis sighed. "It's so NT."

"You know," Mark went on, "I'm an ex-pat waiting to happen. One day I'm going to move to Hawaii. I'm going to move there definitively."

"You should," said Dennis.

"I will. When I have enough dough, I'll move there definitively."

Mark explained that he was obsessed with Hawaii. In his spare time, his passion was to write a regular column for an Internet magazine called *newshawaii.com*. The "mainland," on the other hand, was clearly exasperating for him. It was staid, stiflingly inflexible. When he was younger, Mark's mother had suggested to him that swimming with dolphins might be beneficial to his mental and physical health. "So," he said, "I found a woman in West Oahu who takes autistic people out for swims. She calls us 'unique beings.' So I went. It was the best experience of my life." Then there was Hawaii itself. Its volcanic dramas, prehistoric greenness, and sense of oceanic isolation. It possessed the same relationship to American normality as antimatter to matter, and in a way it mirrored Mark's own relationship to other people: an island tenuously connected to a large mass. His own account of himself was very simple.

"When I was at school I was doing too well to be called mentally retarded. So they called me 'emotionally disturbed' instead. That was the trendy label at the time. Naturally, it's meaningless—I mean, I assume it's meaningless. But it makes NTs feel better about themselves. I wouldn't say *disturbed*. What does 'disturbed' mean? Dislocated, more like. Except when I'm on Hawaii, of course."

On the Islands, he said, he felt free. Unlike repressively gray mainlanders, the carefree islanders in their happily gaudy shirts tolerated one's peculiarities: they let the autism flow.

"So you see," Mark repeated, "I'm an ex-pat waiting to happen."

As I was talking quietly to Mark, I was thinking of Sacks's portrait of Temple Grandin and Asperger's vignette of Fritz V. There seemed little similarity between the three until Mark began to describe his love life. Relationships were a difficult subject. Women eluded him; he could not establish a language with them. With them, he said, he felt like a character in a *Monty Python* sketch—and Mark assured me that he was a huge fan of *Monty Python*. Flirting with women was like the sketch of the man who pays an agency to have a properly structured argument. He acted it out himself:

"I'm here to have an argument. No, you're not. Yes, I am. No, you're not. And so on. I never know what's ironic and what isn't. And I never know when it's a real argument or a pretend one."

"It's not much different for neurotypicals."

"That's what they tell me, but I have a hard time believing it. I used to be a big Deadhead at college, and I must have been the only person in history to leave a Dead concert alone."

This was a glum fact, and I had to admit that it was glum.

"And at Yale?" I ventured.

"A disaster. We Aspies aren't cut out for it."

Mark was the first Asperger person I met. His descriptions of the "mainland," of course, seemed a little paranoid, but with time I began to wonder if they didn't have a persistent logic to them. It's a question of whether one can exist in a culture while in some way electing to simultaneously live outside it. The imaginative life (which in Asperger people is so different from our own, if it even exists at all anymore) can be lived in a spirit of both antic gaiety and hermetic isolation, and it can end up being a kind of alternative planet—in much the way that obsessive fans of bullfighting often describe themselves as living on the "planet of the bulls," or we ourselves will describe misfits as living on "another world." What intrigues us, however, is not that certain individuals can seem to inhabit spiritual worlds that are not our own, or even faintly familiar to us, but that these "planets" have their own gravitational laws, their own chemical atmospheres, and their own metaphorical climates. They are rarely empty, randomly chaotic, or sterile. On the contrary—they are still intensely human and therefore *lawful*.

Darius McCollum, Fritz V., James Jones, Temple Grandin, Mark Ramoser: none of them are especially similar, but a subtle psychological law connects them. There is something in their solitude that cries out for an interpreter of dreams to piece them together—not merely scientifically, but sympathetically.

After the meeting, I went down to Amsterdam Avenue in a relaxed mood. A moist heat dripped from the little soiled trees, and their leaves shook with the reverberations of a construction crew toiling pointlessly on a decayed sewage pipe, as they do every summer. The usual mind-splitting cacophony of Manhattan, which somehow never has anything to do with the material progress that we all vaunt as being its excuse. On perpendicular streets, meanwhile, I could see the Asperger parents melting away into the dark, hurrying back to their lives. They at least were trotting merrily along and, unlike me, they didn't skip awkwardly to avoid the cracks in the sidewalk. Then suddenly, quite unexpectedly, I began to miss my talking with Mark and Dennis. Something had drawn me to them. Where were they now, I wondered, even now, only five minutes after our separation? Did they know how to use the subway or hail a taxi? Did they know how to count money and use maps? Did they have apartments to return to with normal keys and locks? How would they spend the rest of the evening, alone as they always were? I thought of them trying to cook for themselves, playing Scrabble with themselves, or sinking themselves into labyrinthine math puzzles. True, there are television and computers, those antiseptic salves of the solitary. But I was curious to know if Mark ever had the itch to go dancing in a salsa club or wander into the Met's Temple of Dendur at night, or, for that matter, whether he ever fantasized about highjacking the E train to the World Trade Center on a Tuesday night. In the final analysis, he had been impassively inscrutable.

All the way home, therefore, I wondered to myself what kind of child Mark had been. Had he been like Fritz or like myself? Had he been fractious, malicious, or pedantic, or had he obeyed every rule to the letter? And I then recalled that when Darius McCollum's mother was asked what she remembered most about Darius's childhood, she had replied that another boy had stabbed him at school so badly that during class he had to sit with his back to a wall—in other words, at the back of the class. It was clear, of course, that the riddle of the Mindblind lay in their childhoods. For it was as children that they first developed their special sense of apartness and separation: the pursuit of the mystery of Asperger's, therefore, would not only be a quest for the elusive nature of normality, it would simultaneously be a pursuit of the mystery of childhood.

CHAPTER 2

LITTLE
PROFESSORS

"We had the best of educations—in fact, we went to school every day—"
"I've been to a day-school too," said Alice; "you needn't be so proud as all
that." "With extras?" asked the Mock Turtle a little anxiously.
—Lewis Carroll: *Alice's Adventures in Wonderland*

The Hilton Airport hotel in St. Louis stands in a tangle of roads which,
seen from the air, resembles a half-nelson knot. There are cul-de-sacs,
service roads, tarmac arteries feeding Sizzlers and International Houses
of Pancakes, Airport Hiltons, and Red Roof Inns. You cannot help fret-
ting about it. *How*, you think to yourself, *can I possibly get out of here?*
What if you never get out of here? Where will you go to eat in the
morning if you choose not to eat in the hotel? If you try to negotiate the
roads you will be killed instantly; you are trapped, trapped in the St.
Louis Airport Hilton with nowhere to go.

The upper rooms of the St. Louis Airport Hilton look out over the
runways of the Lindbergh Airport, but because the windows are so effi-
ciently glazed, the sound of the engines can't penetrate the rooms. Here
I sat for a considerable time watching America West jets taking off and
landing in a dumb show. I began counting them, noting their numbers,
model makes, and tail designs. Thunderclouds blanketed the takeoff
zone, and the planes vanished into them with a kind of Third Reich
aplomb. Then I remembered where I was. Sit in a Hilton Airport room
for a while and you realize that the modern American hotel has successfully formulated the desires of its clientele who must in their sub-

conscious be looking forward to a grandiose old people's home modeled on a utopian nineteenth-century prison: there is perfect symmetry in everything. Every corridor looks like every other corridor, every room like every other room. Marble panels, suspicious hushing Persian carpetry, droplet chandeliers, plastic magnetic card keys, iris-motif bedcover, picture of a field of poppies over the TV, coffeemaker, low-slung furniture set, sterilized plastic water cups.

As you step into your allotted corridor, you find yourself looking anxiously over your shoulder down a vast vista of smoke detectors, strip lights, and numbered doors. It is, in reality, a kind of mathematical puzzle, albeit one that is easy to solve. Out of curiosity, I often find myself taking the elevators to floors where I have no business being. I walk around, finding that the seventh floor is in every detail identical to the fifth where I am staying—as if this should not, could not be. Even the decorative bowls of silk flowers are the same: same ratio of artificial peonies to artificial tulips, same number of yellow carnations per perforated pot. It is an astonishing performance. In part, an Aspergerish feat.

Here, however, a book publishing company from Texas called Future Horizons was holding its 2001 Asperger's and Autism Conference. Future Horizons is actually a versatile media company, which, from humble beginnings, has grown into a booming business—the brainchild of its CEO R. Wayne Gilpin, former Director of the Autism Society of America and father of a seventeen-year-old autistic son. Indeed, Wayne's son Alex was his reason for creating the company in the first place. The spirit of American self-help is inextinguishable, and Future Horizons is in some ways a distant progeny of grassroots organizations like Alcoholics Anonymous. Today, it bills itself as America's leading publisher on autism and especially Asperger Syndrome, and its conference is a prestigious event—*the* annual national gathering that attracts the "cream of the field." And while I cannot say why this estimable enterprise dedicated to the mysteries of Asperger's had chosen the Hilton Airport Hotel as its venue, the feeling of being inside an airport was actually undeniably delicious—for I believe I have already mentioned the pull that airports exert upon me.

Downstairs in the lobby, the bubbly Future Horizons staffers were giving out nametags to an equally bubbly crowd of teachers, psychologists, occupational therapists, and speech tutors. It was rather like a marketing conference—at least what I imagine a marketing conference to be like. The attendees had gathered from all points of the nation to hear a three-day feast of lectures and seminars given by some of the top people in what could be called the "Asperger's industry." There was Elizabeth Gerlach on "Just This Side of Normal," noted author Carol Grey speaking on the subject of "Confront, Concede or Teach," and Dr. Liane Willey on "Pretending to Be Normal."

It reminded me how popular conferences have become in every sphere of American life. There is no profession or interest group, however humble or obscure, that doesn't have its own meetings circuit. From Ido ("Reformed Esperanto") universal language jamborees to annual gatherings of the Association of State and Territorial Waste Management Officials, the special-interest conference, the gathering place of hopeful idealists or hardened realists, is a linchpin of today's professional life. Each conference circuit has its stars and heroes, whom no one in the outside population has ever heard of. Who is America's greatest conference star in the field of canine-grooming materials? I have no idea, but I'll bet anyone that he or she surely exists.

In the lobby, meanwhile, I quickly found Wayne Gilpin. Wayne is a dry fellow, but likeably fatalistic in the way that many unstinting family men are.

"Look at all the folks," he said quietly in my ear, "it's packed. Are you going to hear 'It's a Long Haul, Wear Sensible Shoes'? It'll be terrific." This, it turned out, was Elizabeth King Gerlach's lecture that day; naturally I said that I would.

"Good, good. Get a seat early. She's quite a star, you know."

He was an amiable man, and I felt more at ease. Since I had an hour to kill before "It's a Long Haul, Wear Sensible Shoes," however, I went to get myself a gin and tonic. The Hilton bar was crowded, and I peered around to see if there were any Aspies to chat up. There were none, though others were clearly neurotypical delegates. A school social worker from Lawrence, Kansas, told me that among her children "anxiety and depression," rather than subtler forms of autism, were the norm. With no expression of surprise or bewilderment, she explained that the mental problems of children these days were indistinguishable from those of adults. And that they were treated in more or less the same way—that is to say, with the same psychotropic drugs.

"Doesn't that alarm you?" I asked.

"Oh, no. We find Ritalin works perfectly for them. Makes them really calmer."

"And what if you found an Asperger child?"

"We already have three of them. We're thinking of using Auditory Integration Training. That's AIT, you know."

It was better not to ask what AIT was, so I nodded.

"AIT," she smiled, "makes them really calmer."

Meanwhile the bar was alive with what could be called psychiatric conversation. A woman behind us suddenly piped up, "Anyone know what NLD is?" The question shot back and forth across the room in high voices.

"Neurolinguistic Disorder?"

"Don't know, honey."

"Nonlinguistic Dysfunction?"

"Whatever it is, honey, I've been diagnosed with it!"

The Hilton lecture hall was indeed packed. Elizabeth King Gerlach was quite a beautiful woman, a young blond from Oregon whose son Nicky had been diagnosed with Asperger's when he was seven. She stepped up to the podium and launched breezily into her lecture.

When he was seven, she began, Nicky was "out of control." He would spin all the time, throwing himself around like a human top. When he was asked what he wanted to do when he grew up, his reply was, "I want to be a screwdriver!" Not a fireman? Elizabeth would ask him. "No, a screwdriver." A human screwdriver. He was plagued by ear infections and sleeping difficulties, as well as having tantrums. Elizabeth took him to a therapist, naturally; the latter asked if he spun plates. No, she said, he didn't, but he did like spinning and lining up his toys as well as rushing around throwing birdseed everywhere. The prognosis of the therapist was swift:

"Well, it was Oppositional Defiance Disorder."

The audience sighed knowingly, and heads turned to each other with equally knowing nods.

"But I wasn't satisfied," Elizabeth continued. "So I decided it was autism. I found a diagnostic team at the University of Oregon to take the matter forward."

What made the team at the University of Oregon conclude that Nicky had AS? In part, it was his obsession with ships. Nicky loved ships so much that he turned his bedroom into one. On the projection screen, Elizabeth summoned up a picture of him, a slender fair-haired elf in a yellow hat sitting in his nautical room with a huge "sail" suspended above him. On the sail, a fish had been painted. He read *The Great Age of Sail* and sang the same songs over and over: "Sailing the Ocean Blue" and "What Shall We Do with the Drunken Sailor?" He would spend much of his day building Lego ships, then destroying them. In all, he looked like an interesting kid, and I immediately wanted to meet him. His poetry was distinctly intense:

> I hear the beat
> It's inside my head
> I always hear the beat
> Can't stop the beat
> It's always in my head

I wondered if he wrote similarly hypnotic stanzas about ships' rigging, nautical steering wheels, or the bad habits of Bluebeard.

But if Nicky appeared to be intense, the treatments applied to him were even more so. He was taken off sugar, wheat products, and cow's milk; given vitamin B6, magnesium, chiropractics, foot massages, therapeutic horseback riding, antifungals, Prozac, Valium, and doses of Auditory Integration Training. Following a thing called the Dan Protocol, a biological diagnostic test set up by the Autism Research Institute, his stools were analyzed for signs of yeast overgrowths and urinary peptides. The biological basis of autism, Elizabeth explained, could be seen in the metabolic areas; autism must therefore be chemically curable, just as cancer one day may be—even if the Prozac had backfired in Nicky's case.

A picture of a child dialing inside a telephone booth had appeared on the screen. Dialing 911, of course. Elizabeth shrugged.

"Prozac works wonders for me. It helps me contain my obsessive thoughts. But it sent Nicky to the moon."

By now we knew that Asperger Syndrome was indeed a long haul. Of course, Asperger's was biological, not developmental. The bad old days of blaming Bruno Bettelheim's refrigerator mothers were long gone. A kind of low gasp of contempt swept across the audience as the word "Bettelheim" was spoken, and I was glad he was not present—or even alive. I also began to feel a little uncomfortable. Might some iconoclastic colleague have written a blunt investigative exposé called *Prozac Mothers*? I very much hoped not. Moreover, Elizabeth's language had taken a serious turn. She was now explaining the Eight Elements of Effectiveness and the Early Intervention Program. There were lists of things to do and sublists of lists of things to do. "Initiate functioning interaction," she told her audience, "develop self-monitoring and self-management skills, create predictability."

It was clear that Elizabeth was speaking from cruel experience— the day-to-day heart-wrenching slog involved in enticing a dysfunctional child to behave. The key, presumably, is to create order and routine such that the child can follow it mechanically. But at the same time I couldn't help wondering at the language the conference attendees seemed to share. It appeared essentially to be a corporate lingo whose vocabulary was relentlessly technical. It contained phrases like, "the use of augmentative communication methods," "data-based management programs to track the child's progress," "functionality," "specialists," "programs," "models," "interactions," "sensory integration input modality," and so on. It dawned on me that today this is the accepted language of parents in dealing with their children: parenting as benevolent engineering. The language itself has become universal and smoothly internalized, unconscious almost, and the child is seen simply as a machine that has gone wrong.

It is a language that is also characterized by euphemism, so much so that euphemism is now almost synonymous with normality. I remember Rudolph Giuliani a few weeks after the World Trade Center atrocities talking to a group of schoolchildren in downtown Manhattan who had complained of their terror and the awful smell of death in the air. "What you children are going through," our usually blunt-tongued former mayor said to them as he held his arms open wide, "is a wonderful learning experience."

At the same time, a terrifying question must always confront the parent of an Asperger child: What exactly are you *supposed* to do? It's understandable, then, that people should go looking for a working language into which they can translate both their terror and their hope, a language that stubbornly asserts their optimism for the futures of their children. Despite my misgivings, I had to admit that I found it admirable.

Elizabeth actually spiced her lecture with a phrase she had seen the day before in an American Airlines brochure. "It's important for good people to continue the good fight." There was wild applause. Then something very strange happened. Reaching below the podium, Elizabeth hauled up a Groucho Marx mask and put it on her face. From behind the plastic nose, glasses, and exaggerated moustache, she exhorted her audience to have fun at their next school meeting. They should be subversives, play Frank Zappa in the background, create havoc among the complacent school administrators. In short, they should wear Groucho Marx masks and be like children.

"After all," Elizabeth concluded triumphantly, still in her Groucho guise, "Asperger's is everywhere around us. You know, I think my uncle John is AS. I'm sure of it—he's really, really quiet. He never says a word. And he just loves windmills. He *loves* them."

★

At lunch hour the next day I drove into St. Louis out of curiosity. The only part of the city I like is Soulard, the old French quarter with its little brick houses, rural-looking pubs, and crooked sidewalks with maple trees. I tried to remember that this was the city where T. S. Eliot had come from, and from which he'd fled, but the spirit of Prufrock belongs more in London, and the exquisitely American Greek-literate dandies of the 1890s are a long-departed curiosity. I wandered around the covered market listening to refugees from Serbia, Montenegro, and Kosovo discussing cabbages and the price of grapes in their own languages, and went for a winding walk through the hauntingly nostalgic grid of nineteenth-century lanes around it. I kept wondering why there were no children to be seen. Downtown is the same: lunchtime in spring and not a child in sight among the loopy glass towers and Gropius-like boxes. Our architecture itself has made the rambunctious presence of small children inappropriate, if not downright impossible.

After lunch I went to see a woman from Texas named DeAnn Hyatt-Foley, who delivered a lecture entitled "How to Be Your Child's Social Skills Coach." DeAnn began by telling us that her son, Matt, had read Liane Willey's book on Asperger's, *Pretending to be Normal*, and had promptly announced, "I have this!" Now this got DeAnn thinking. She became more and more certain that her husband Ryan had it, too. Not only that, but it became all too clear to her that she herself has certain aspects of AS. Before she knew it, she had come to the realization that they were, in fact, an entire Asperger family. The symptoms seemed familiar enough: stiff posture, lack of facial expressions, a distinct style of dress, dishevelment. Consulting a thing called the Vineland Adaptive Behavior Scale, DeAnn realized that they were not at all normal.

But here there was a difficulty. "Where," DeAnn asked, "does the Asperger Syndrome begin and the child end, and vice versa?" To solve this very real conundrum, the parent must follow endless rules of engagement, as well as absorbing advice, using various systems, programs, clauses, scripting, role playing, and perhaps even a bit more equine therapy. I wondered if any parent would ever be able to remember even half the suggested exercises, programs, routines. It reminded me of the way in which children themselves are programmed to the point of exhaustion in American schools, harried from activity to activity without any intervening unscripted pause, without any time to gather themselves.

I had begun to feel a strange aura of gloom at the thought of all these therapeutic regimes when DeAnn herself brought up a quiz on the screen. It was a scripted moment of play. The quiz was about Texas. Which, the quiz read, is the least likely thing to happen in Texas? a) a white Christmas, b) a tornado, c) winning the Texas lottery. The adults assembled all began to hoot and raise their hands like a classroom of kids. A few minutes later, after a discussion of the Theory of Mind, another quiz popped up. What will an armadillo do if you see him? The audience erupted in giggles and calls. It seemed that they had already forgotten the Vineland Adaptive Behavior Scale and were psychically concentrated on these Texas quizzes. Nonplussed, I slipped out and into the bar next door. "My son," I heard DeAnn declaim from behind closed doors, "is obsessed with oscillating fans!"

That night I joined a few of the speakers at a dinner thrown by Wayne Gilpin at the Kreis's steakhouse on Lindbergh Boulevard in south St. Louis. I drove with Wayne, his daughter, his teenage autistic son Alex, and Alex's autistic friend Rob. The boys made laborious puns as we went, using TV commercials and passing neon signs as material: the Hilton became the Mountainton, the highway became the low way, and so on. Alex and Rob were more on Kanner's side of the fence rather than Asperger's; they did not have the nervous, fluttering nimbleness that I had now come to recognize in Asperger people. They laughed uproariously and sometimes uncontrollably at their puns while the rest of us smiled politely. The warm bond between father and son, however, was touching; no interactive communication treatment modalities here. They spoke to each other normally, with direct affection. The boy had developed, it was clear, a supple warm relation to his father that was palpable, reassuring.

Kreis's is a shrine of Midwestern gastronomic luxury, with its cave-like dark interior, poker-hall lampshades, aproned waiters, and imitation impressionist paintings in boisterous gold frames on the walls. The fare was extravagantly bloody, but preceded by odd little bleached salads drowned in viscous sauces. I was seated with Carol Gray, DeAnn Hyatt-Foley, and a palely blond woman who told me that she was the mother of two Asperger boys. She promptly produced pictures of them, two adorable little imps. How, I asked, did she know they had Asperger's?

"They can't read other people's minds. And they perseverate."

It was an odd word. I needed a translation. "They obsess?"

"They go on and on about a single subject that interests them. We call that perseverating. Of course, that could just be a male trait."

"Well," I said, "*is* that just a male trait? Can you read other people's minds?"

"No, but I can read body language. My husband has trouble reading body language."

"So does he have Asperger's?"

"Not really—"

She thought hard, her fingers crossed. "No," she repeated, "not really."

It was a jolly meal and I got fairly drunk on the house margaritas. At one point Wayne blurted out something quite interesting. "A lot of these diagnoses of Asperger's," he said bluntly, "are just completely off the wall, aren't they?"

The women protested.

"There are just a lot of gray areas, that's all," said Carol Gray.

"I didn't mean all of them," Wayne added quickly. "Just some of them. Some of them are just so vague. They're not specific."

"But that," said DeAnn Hyatt-Foley, "is the nature of all new syndromes. We're just getting started here."

"I just have this feeling," Wayne persisted, "that people are some-times being misdiagnosed. . . ."

The women huffed and puffed. No, no, that wasn't it at all. They declared that some misdiagnosing was inevitable with a new syndrome. Far more significant, they pointed out, was that thousands, perhaps millions, of people had gone *un*diagnosed.

"There may be millions of them out there," said the pale mother of two. "I wonder about my own relatives. A lot of them perseverate."

I thought about myself: Did I perseverate ever? I was sure that I did, and that it was something of a vice. I surely perseverated about *Iron Chef* and other such idiotic things.

"I know I'm half Asperger," DeAnn said breezily, with complete con-fidence. "I must have at least three or four symptoms. What about you?"

"I have at least one," I laughed back. "I tend to perseverate."

"Ah!"

But suddenly they were not laughing. They were nodding with serious compassion.

"Do you? Well, if you perseverate," Carol Gray said, "perhaps you should go see your doctor."

I enjoy perseverating, I thought.

"Yes," I said, "that's an idea."

On Lindbergh Boulevard on the way back into St. Louis, Wayne told me how delighted he was with the high turnout of the conference. It had been booked out weeks in advance. Asperger Syndrome was all the rage these days; people were flocking to a syndrome for which there was not yet a scientific diagnosis. It seemed to explain so much to so many people: their eccentricities, their feelings of alienation, their tics and obsessions. It had become almost a barometer of the American psyche.

I wondered if this might be true, at least for some. Could a real medical condition give shape and substance to an unreal one, to a vague cultural feeling of un-ease or dis-ease? Or was it possible that a broad and growing cultural phenomenon was finally legitimizing a disease that heretofore had been really just an exotic rarity?

Then, floating by in the dark amid the landscaped communities and vast low-density malls, I saw a forbidding sign rising up on a small grassy knoll. It was the HQ of Monsanto, the multinational chemical and agribusiness and biotech company. Under the word Monsanto itself, three other words were inscribed in the stone with an imperially Roman simplicity: Life, Health, Hope. I waited for Alex and Rob to make a bad pun, but none came. For some reason, this put me into one of my somber moods in which, interestingly enough, I tend to obsess and perseverate. Furthermore, I was brought to remember a strange refrain in a poem by an English Asperger's sufferer and writer named David Miedzianik, who is a frequent visitor to the annual St. Louis event. His poem, which I had found by accident a few weeks earlier, was entitled *At the Autism Conference in St. Louis 1997*, and it was now all too pertinent to my confusions. Miedzianik wrote:

> Well I was at The Autism Symposium in St Louis again.
> Some of the things they were talking about there were too hard
>   for my brain.

Back at the Hilton, therefore, I went up straight to my room intensely anxious to catch the latest installment of *Iron Chef* live from Tokyo. Luckily, I was just in time to see Bobby Flay, famed chef of Mesa Grill, defeat Iron Chef Morimoto in the Battle of the Japanese Lobsters.

★

To wax diagnostical about the mood swings of entire societies is always perilous. America is not a monolith, medically or any other way. Nor are its disorders unique to itself, for in the Western world there are always mutual cross-infections both benevolent and noxious, and there is often little distinction between European and American neuroses. And some afflictions seem to travel well beyond "the West." Eating disorders such as anorexia, for example, have spread around the globe from the industrialized nations to the formerly inaccessible rural interiors of societies like China and India. These disorders seem to follow in the wake of cable and satellite TV, as do newly experienced forms of depression. But when it comes to what is dubiously called mental health, America does hold a special place, with its extraordinary institutional skeleton, a now huge and complex and self-perpetuating apparatus, that seems to support an ever-growing array of psychiatric disorders of its own making, which as they grow express a peculiarly American notion of personality.

Nor is it rash to point out that in America, psychiatry has assumed an unusual importance—one that is gradually supplanting other and earlier authorities not just to treat, but to name and categorize and even, it must be said, invent illnesses of the spirit. The therapist is surely the contemporary version of the confessional priest who measures with some elusive instrument the fluctuating states of his flock's tormented souls. Although therapy is of course not the same as psychiatry, the former's ubiquity in its Hydra-like multiplicity is nevertheless an out-come of the rise of psychiatry's prestige and its acquisition of an author-ity, which now far surpasses the cloak-and-dagger glamour once enjoyed by Freudian analysis. Psychiatry today is America's secular reli-gion. And in the case of people who actually need it, which especially includes people like the Mindblind, its extraordinary cultural prestige might actually be a double-edged sword.

For, strangely enough, therapy is what most binds the Mindblind to the emotion language of mainstream culture. Through it, the latter subtly enters their closed world and alters it. This is not only because of the inevitable nature of currently fashionable treatments. It is also because therapy itself often implies a particular view of childhood, though not always one that is explicit.

From that same Austrian import of the early twentieth century, today's therapies (which in their practically infinite variety have ceased to be anything very precise) retain at their heart a pursuit of a cure for the wounded inner child. The drug culture of the 1960s, which the Baby Boomers regard as their Paradise Lost, has quietly given this generation a more respectable means of indulging itself in its neurotic moods: the antidepressant, which bears an unexamined similarity to the substances with which it deranged itself thirty years ago. Looking for the child in themselves, the 1960s generation that now rules the nation has had great success in finding the hurt child within, but what of its own off-spring? Their kids do not work in salt mines, they are hardly Oliver Twists. But they most certainly are the progeny of a therapeutic culture that has duly colonized them as successfully and as ruthlessly as Cecil Rhodes did when he took possession of the Transvaal.

The collapse of traditional cultures has meant the collapse of tradi-tional therapies, which normalized people by taking them out of their purely private neuroses and immersing them in the vast experience of other generations. Psychiatry, which now regulates the inner life, has no sense of historical time—it operates only in the present moment, or at most in the time frame of an individual life. Thus, it cannot take people out of themselves in a profound way; it can only make them dance around themselves in an eternally shallow waltz. As it does so, the "diseases" it can recognize begin to run amok, as if they have a life of their own.

According to a report published by the Centers for Disease Control and Prevention in May of 2002, 1.6 million elementary-age children had been diagnosed with some form of ADHD—Attention Deficit Hyperactivity Disorder—by the end of 1998. Parents of 7 percent of all children reported that a doctor had accordingly diagnosed them as having the dread syndrome. The National Institutes of Mental Health, meanwhile, estimates that 4 million children have some form of learning disability, 20 percent of them associated with attention deficit. The NIMH also claims that one in ten American children suffers from some form of "mental illness," mostly depression, ADHD, or "emotional disturbance," while the American Academy of Pediatrics tells us that at any given moment one in five of them suffer from what it calls "mental health problems."

As the number of children diagnosed with such problems has increased, so has the use of psychotropic drugs. The National Institutes of Health reports that by 1995, 6 million prescriptions for stimulants such as methylphenidate (Ritalin) and Dexedrine were being written every year, and that between 1991 and 1995 there was a 2.5-fold increase in the use of Ritalin alone. Lawrence Diller has written that "an increase of this magnitude in the use of a single medication is unprecedented for a drug that is treated as a controlled substance." This suggests a problem, he adds, of "epidemic proportions." Diller notes that in 1990, in his California practice, he needed a pad of one hundred Ritalin prescription forms every nine months, whereas by 2000 he needed one every three months.

Moreover, a paper published in February 2000 in the *Journal of the American Medical Association* states that the use of psychotropic medications among two- to four-year-olds had increased 30 percent in the same period. Such prescribing of drugs for toddlers is known charmingly in current practice as "off-label," because there has been no controlled study of the effects of such drugs on children so young. The NIH, however, does not dismiss the idea of their use altogether. Far from it. Instead, the institute frequently refers to its own recent study of

600 children aged from seven to nine who were treated with various psychotropic drugs over a period of fourteen months. This now-famous investigation, the results of which were published in the *Archives of General Psychiatry* in December 1999, has become a benchmark reference for those who believe that seven-year-old children can be effectively treated in this way. Results, say the report, were generally positive and safe. "Off-label" prescriptions might be dangerous and lead us into unknown waters, the NIH admits, but then so are the futures of disturbed children who remain untreated. For this reason the NIH now has seven Research Units in what it calls Pediatric Psychopharmacology. The feeling would appear to be that we are on the threshold of a bright new scientific era of subtly effective treatments. But in psychiatry, of course, we are always on the brink of a bright new scientific era.

Many feel that there has been a collapsing of the way in which children and adults are viewed by the psychiatric establishment. For just as there has been a steep rise in the number of children claimed to have ADD, there has been a corresponding rise in the number of adults claiming to suffer from exactly the same disorder. In part, the rise in medications among children reflects a crisis in the schools. The contemporary middle-class child is very unlike his or her privileged Victorian or early to mid-twentieth-century ancestor. There was a time, and it was not so long ago, that children, at least children of the upper class, were creatures endowed with plenty of leisure time, and childish apartness was actually exalted. Today's children, in contrast, are harassed creatures, whisked through punishing timetables in schools and pressed into a stifling conformity of socialization, participation, and group cooperation. The exceptional child, the child who perhaps prefers not to join in, is almost inevitably seen as problematic, if not downright disturbed (and disturbing).

Diller also wonders if the regimenting of children has led to a crisis of their beyond-the-norm individualism. "Is there still a place" he writes, "for childhood in the anxious, downsizing America of the late 1990s? What if Tom Sawyer or Huckleberry Finn were to walk into my office tomorrow? Tom's indifference to schooling and Huck's 'oppositional' behavior would surely have been cause for concern. Would I prescribe Ritalin for them too?"

Drugs, then, are more the symptom of the problem than its actual cause. As children are driven into abnormality by the pressures of an onerous *normality*, so adults try to placate them by any means possible.

At the same time, new disorders are legitimized by ascribing them to endless numbers of illustrious figures. As with Asperger's, there is a minor cottage industry in diagnosing illustrious persons as having had "attention deficit." There are Beethoven, Ben Franklin, John Kennedy and various other U.S. presidents, not to mention the inevitable Einstein. One ADD expert even described Bill Clinton as being "one pill away from greatness." (As yet, no Béla Bartók.)

Part of what we are seeing may be a set of culture-bound syndromes intricately related to a complex historical change: the hemming-in and consequent decline of childhood itself. It's a claim most professional doctors and psychiatrists would scorn out of hand. But it could also be argued that their own failures and cumbersome confusions merit that we look at mental health in a wider context. Furthermore, though these issues may not directly explain Asperger Syndrome (because the latter is indeed a neurological disorder), they surely have a bearing on the way those suffering from the syndrome are socialized. If childhood normality is confined to progressively narrower definitions, what effect will this have on children themselves, normal or otherwise? What effect, for that matter, would it have on adults?

When I read the papers of Hans Asperger I am struck by his veneration for children, the clear separation of himself from the inner world of the child, which is seen as sacrosanctly different, if not categorically unknowable in many ways. He steps around them with a hesitant and unassuming fascination, but also not without a gentle presumption of authority. What a change a half century has wrought. The dialectic of adult and child has broken down. As the culture infantilizes itself and adult consciousness diminishes, we turn, as children are wont to do, to the Doctor and his multicolored remedies. Is it possible that children, with their increasing crises, are themselves simply responding to the end of childhood? And is psychiatry, like the criminal justice system, finding it more expedient to conceive of children in the medieval way— that is, as miniature adults capable of adult crimes? We have indeed come full circle if children are no longer seen as sacred—as being, as

Neil Postman beautifully puts it, "the living messages we send to a time we will not see." Then again, neither do we see as sacred the adults who now sinisterly mirror them.

★

In children whose relations with the surrounding world are disturbed in some way, the issues of too-restrictive definitions of normality become intensified. But it can be a paradoxical intensification. Autistic children, like deaf children, are in some ways cut off from the cacophony of our culture's media and exist mostly within themselves, in what's been called a kind of "purity." They can appear to us like islands of serenity and sanity. Above all, they make us think about the sheer differentness of childhood itself. What is a child, and how should a child speak and feel? Does childhood possess a kind of mysterious autonomy that we do not yet understand? Does childhood itself have a language?

When I visited my first Asperger school, in New York City, I wasn't sure what kind of children I would find there, or even if they would be children in the ordinary sense at all. Would they be, like the untamed mutes of François Truffaut's film *The Wild Child*, isolated by their failure to master the social language of the majority? Or would they be like creatures out of *Alice's Adventures in Wonderland*? By now I had met some Asperger adults who seemed to have learned the tricks of social language. But children are a different matter entirely. With them, everything is laid bare. In them, a disorder cannot hide beneath the wrapping provided by a subsequent education. As it turned out, I was to be surprised on all counts.

The STAR program is part of a charitable foundation called the New York League For Early Learning, which runs a school for Asperger children, a half-dozen cramped rooms halfway up a narrow building on West 19th Street. On my visit in the spring of 2001, I was met at the elevator by its director, Jeannie Angus, one of those stiffly resilient woman principals I remember from my own school days. She asked me to glide around the classrooms as quietly as possible: unexpected noise upsets Asperger children. "We teach them everything," she whispered into my ear, as we crept into the main classroom to watch five six-year-old boys. "We teach them that when people ask, 'How do you do?' the verb 'do' is just a manner of speaking, a metaphor. Of course, they have a terrible problem with that. They always say back, 'Do *what*?' We have to tell them, it's a kind of play, a way of pretending. We have to laboriously explain to them that words don't always mean what they're supposed to mean."

The boys were seated around a circular table drawing blue whales with crayons. Today's subject was "The Whale Who Got Stranded on the Beach." One bespectacled tot, Asa, was drawing his whale with the meticulous fussiness of an old jeweler setting a stubborn diamond. He had scrawled the word "wale" underneath. The room was filled with sunlit bean chairs, shelves of books like *The Great Kapok Tree* and *The New Club Hut*, a board with a vertical line of words that read *prudence, pretzel, prairie, purple* and next to it the cryptic sentence "Parrots live here, it gets dark and the insects come out."

I noticed that around the rim of the table the outlines of hands had been painted in different colors. The boys were expected, when asked, to control the motions of their hands by placing them over these silhouettes and remaining still. This motionlessness is enforced from time to time by the teacher, who asks each boy how well he thinks he has been sitting still. Each boy then gives himself a grade from one to three, three being the top.

"How was your sitting, Aster?"

"Three."

"Nice sitting, Aster. Were you polite, Henry?"

"Three," from Henry.

"Good work, Henry. How about focusing, Jean-Paul?"

"Three."

"Nice focusing, Jean-Paul. And what about looking in the eye, Asa? Did you look in the eye?"

"Three."

"Well, nice eye contact, Asa!"

They are then asked to put their hands on the silhouettes and hold them there for a few moments. Eye contact, hand control, focusing, and manners: the attributes of external normality, which Asperger children have to learn by rote and by example, almost as one learns a difficult foreign language. It is clearly a punishing routine for these young teachers, who have to patrol not only a child's normal disobediences but also to iron out errant tics, drifting eyes, lapses of motor control, and the resurfacing of monomaniacal obsessions that often explode in the classroom with alarming force.

I noticed also that none of the boys had looked me in the eye, but that they had immediately noticed a Babar the Elephant pen that my girlfriend had picked up for me at the Guggenheim Museum gift store. It had a Babar head mounted on a flexible spring, a gimmick expressly made to invite twiddling. One by one, the boys broke away from the communal table and drifted over to me to make their inquiries about this elephant head that I was flicking back and forth with my thumb. Interestingly, as they did this, not one of them actually looked me in the face. Instead, their eyes were always directed elsewhere. The effect was like that of being stared at by someone who is cross-eyed.

"Where did you get that?"

"At the Guggenheim."

"Is it French? Why is it prehensile? He must have a sore neck."

I was sure I had never heard a six-year-old use the word "prehensile" before, and certainly not while actually knowing what it meant. But such precocity is not rare with Asperger boys. "They often have sky-high IQs," Jeannie Angus said, as the boys scattered onto the beanbag chairs and rolled around with their feet in the air. "They seem so normal, don't they? But look a little closer, and you see these tiny cracks."

She pointed out an angelic putto drawing by himself in a corner. "He used to come into school and squirm on the floor to get a feel for its texture." Another boy, she added, had been terrified by butterflies during a school visit to the Natural History Museum. But when an old woman came up to him to try and comfort him, he began screaming at the top of his voice, "She's trying to kill me!" I looked at the boy again. He was mouthing words to himself, sedately flopped into a chair, his huge lashes trembling restlessly as he flipped over the pages of a book. Jeannie pointed to yet another student, a slightly larger boy who was standing alone, playing with a stalagmite of play-dough. "Michael knew everything about tornadoes. He thought he *was* a tornado. He whirled around the room breaking everything. Then he would tell you what his velocity was."

Michael also knew the velocity of every famous tornado in history, and was something of an expert on things like G forces and the statistics of tornado-related destruction. "He was five years old and he was like a videocassette about tornadoes—a video that he could rewind and play over and over. He was really happiest when he was acting as a tornado."

A little later, I peered into the room next door and saw a tiny girl with wild *Wuthering Heights* hair kneeling alone on a linoleum floor with a three-meter line of toy dinosaurs in front of her. The perfectly arranged procession seemed to stretch from one side of the room to the other. I had the impression she was the only girl there—a rare creature? She looked up with grave eyes and went on with her dinosaur count, touching each head as she made her way along the phalanx of beasts.

In yet another room, this time closed, Jeannie showed me through a window seven children diagnosed with more severe forms of PDD (Pervasive Developmental Disorders) sitting around a table trying to concentrate on an English lesson, their faces brimming with a kind of bridled energy that appeared to be on the brink of spilling over into violent farce. "You see the difference." She made me look at their heads, which shuddered and twitched from time to time in a way quite unlike those of the Asperger kids. The latter, she said, really did seem to be in a category by themselves. "But why don't you come and watch them learning faces?"

Learning faces is one of the truly startling aspects of Asperger remedial education. In the main room, the boys were now seated obediently in front of an easel, upon which stood a board covered with simplified drawings of human faces. Under the smiling face was the word "happy," under the frowning one, "angry," and so on. A teacher pointed to each face and asked the boys what each expression meant. It appeared that most of them had this exercise down pat. They knew exactly what a smile meant because they had been told so many times that by now they could hardly not know. The simplest syntax of emotion had been demonstrated to them, and little by little they were beginning to adopt its conventions. "One parent told me she thought Asperger boys were counterfeit bills," Jeannie concluded. "But they're bills all the same. We can teach them to pass as normal."

A week later I met Asa with his parents at a Starbucks on East 87th Street. Asa's father, Philippe, also has Asperger's, and I was interested to see how father and son dealt with one another. Asa himself looks like the prototypical "little professor," with his round wire Harry Potter specs and his long words, while Phil, his generational alter ego, is an

aspiring actor who grew up in Berkeley, the sheltered son of a law professor. I couldn't help thinking that Phil was the adult version of Asa, a
matured version of an adorable and precociously dotty boy. Again, the
nervous, bird-like delicacy of gesture, the unsure picking of words and
expressions, and the dry self-deprecation that would have fitted better if
he had been English. Like many Asperger adults, Phil is self-diagnosed
but readily confirms a lifelong state of bemused noncommunication
with the rest of the world. "I studied acting at Harvard," he said at once,
while introducing his estranged wife Lisa, who is Asa's mother. "But my
attempts at camaraderie and constructive criticism were immediately
interpreted as acts of warlike aggression and hostility. I was thrown out,
just like that." He snapped his fingers and grimaced. "It's life as one
long non sequitur. One non sequitur after another. I have nightmares
about non sequiturs."

I went off and got myself a coffee. For some reason I was having
trouble breathing. It must have been the Starbucks itself. The master
planners of Starbucks have understood only the technical appearance of
coziness, not its real nature. Cramped in this cappuccino-serving dollhouse you are left feeling airless, perilously close to a non sequitur. It
was a feeling that Asa understood, because when I asked him if he liked
this place he frowned and shook his head, popping two index fingers
into his ears. "Noisy?" I asked.

"I can't breathe," he said. "By the way, do you like those pictures on
the walls?"

I looked up and saw that they weren't pictures at all but glossy
coffee-table books about *The Cocoa Bean* and *Great Teas of the World*.

"I like those Indian ladies," he stuttered on, "picking tea leaves. Are
they happy?"

"I couldn't say if they're happy. They're smiling at least."

"So are we. But we can't breathe, can we?"

Asperger people rarely marry, but Lisa and Phil had met while
waiting on tables at the Vienna Café in Los Angeles. Lisa always knew
he was a little odd but admits that she was drawn to that very quality. I
thought she looked a little tired, as if her patience had been worn thin
by long trials that may or may not have been connected to her ethereal
ex-husband. Phil, on the other hand, seemed boyishly fresh and alert,
the kind of person who is always bubbling with projects that might
have been minted in his mind only the night before. Currently, he was

working on a long theatrical monologue about the inner trials of Asperger Syndrome. He immediately began an earnest conversation with Asa about their dreams.

"I saw my room collapsing," Phil began enthusiastically.

"That's cool. Were you scared?"

"Very."

"Do you have a lot of nightmares?" I asked Asa.

He blinked, his eyes huge behind the round lenses. "I can't sleep much. My mind doesn't stop. Not ever."

"And what do you want to be when you grow up?"

I hoped he wasn't going to say "screwdriver." But his reply was conventionally ambitious.

"I want," he said, "to be a doctor *and* a dentist."

His heroes, he went on, were Mary Poppins and Roger Rabbit. Roger Rabbit could *dance*.

Almost from the beginning, Lisa noticed that Asa was as odd as her husband. She could never banter with him in the way that a parent does with a child. He was too grown-up in his manner and interests, and, as with an adult, his conversations always had to have a point. Small talk was out of the question. For his part, Phil said he saw much of Glenn Gould in Asa; in fact, the great late Canadian pianist is a touchstone for Phil, the very type of the aloof, transcendent "Asperger genius" of impeccably inscrutable mystery. And Asa, too, is preoccupied with the Canadian virtuoso. For some reason, he refers to him as "Mr. Rathburn." When having minor tiffs with his parents, Asa will sometimes shriek, "I'm going to tell Mr. Rathburn!"—something that the original Mr. Rathburn would have relished immensely.

Asa also has an imaginary friend called Ehe (pronounced "ih"), who is an inventor and scientist. Ehe, in actuality, shares most of Asa's interest in things scientific. Ehe "asks" Phil to buy Asa science videos and shows him how to draw remarkably accurate diagrams of atoms. Asa's insights are sometimes worthy of Ehe, too. The previous night, Phil said, Asa had calmly informed him that written numbers were fast, and that spoken ones were slow.

"That's Asperger loveliness," he added quickly. "Things slip and slide like a saddle on a wild elephant!"

Phil is something of an Asperger activist, believing that these children have their own intellectual vector that should be neither denigrated nor forced back into conformity. Indeed, according to Phil, it is Asperger intelligence, and not that of normals, that holds the keys to the future. The world might even be condemned to intractable crises without the guiding light of nonneurotypicals to correct its zany and dangerous ways. "I believe that very strongly—we are here to save you!" But in Asa's case, the schools he had attended have been predictably reluctant to acknowledge such a gift. They were, instead, a nightmare.

He started out at PS 198 on the Upper East Side. One telltale difficulty was his inability to perform the fire drill. Asa stubbornly refused to recognize the fire drill's logic because it was a form of pretence, of play based on pretence, and AS people cannot understand pretense. They can no more utter lies (or even innocuous social fibs) than they can run around in a fire drill pretending that their school is on fire when it is clearly not on fire at all. Perplexed by this relatively simple lack of metaphoric reasoning, the school suggested that Asa needed therapy, the instant panacea of the age. But what kind of therapy? Since Asa had not been diagnosed with any kind of disorder and because he was not obviously abnormal in any way, the school simply allocated him a posse of therapists: four of them in all. There was a communication therapist, a socialization therapist, a physical therapist, and a speech coach. These four specialists would follow Asa around during his school day and sit with him in his classes. "Needless to say," said Phil wearily, "they made him feel ten times more abnormal than he had before."

"I called them the Four Shadows," Asa put in.

His anxious parents asked the school if they didn't think this was a touch of overkill. Four therapists per child? The school replied that in general the kids "loved the attention."

"It's the culture of conformity in American life," Phil went on. "Especially in American schools. They're like touchy-feely penitentiaries. It's a regime of control, pure and simple."

Lisa: "We were told that he couldn't get into the Lab School for gifted children because he was 'too quiet.' Quiet? Since when is a kid singled out for being *quiet*?"

But of course as well as being quiet, Asa would also alarm his elders by throwing himself on the ground, clasping his hands together, then jump up and down in a seeming fit of ecstasy. It's behavior that the STAR school knows how not to repress.

I asked Asa if he liked to impersonate made-up characters (I'd read this in a book on Glenn Gould, who had many crazy characters up his sleeve). Father and son began chortling—I was going to say "lustily."

"Like my dad," Asa said. "He does that."

"In school I was always inventing a persona for myself. One month I'd be Groucho Marx, then I'd be Sherlock Holmes. I was trying to find a character which people would like, but it never worked. Eventually, *they* called *me* Mr. Peabody, that pedantic little talking dog! That's me!"

"What about Mr. Rathburn?" Asa asked, his eyes now twinkling behind the professorial wire frames.

"He's a great character, isn't he?"

But Asa had now become matter-of-fact. His voice was level, poised, with that deadpan accuracy of the schoolmaster of yore, though not of now. Suddenly I could see why his mother could never have childish conversations with him. And yet in his sense of uncontaminated fantasy he was pure child. "Personally"—the voice hovered in a C flat—"I'd rather be Roger Rabbit. Roger Rabbit can dance. I think Mr. Rathburn and Ehe would agree that Roger Rabbit can dance."

★

One of the best known classification of the different types of Asperger child has been set forth by Lorna Wing, who in essence "rediscovered" Asperger's in 1981. Wing discerns four categories that seem to my ear quite characteristic of all children, autistic or not. She names them as follows: the "Aloof Group," the "Passive Group," the "Active but Odd Group," and lastly the probably pretty melancholy "Over-Formal, Stilted Group."

According to Wing, the Aloof Group is the most common. These children behave as if other people do not exist, and they show no interest in or sympathy for other people's pain. They seem "cut off, in a world of their own, completely absorbed in their own aimless activities." The Passive Group, meanwhile, is the rarest of the four. Children belonging to it cannot initiate social interaction, but at least they have the virtue of being peaceable and compliant. The same is also true for the Active but Odds, but these latter have other problems.

Asperger people are conventionally said to have problems establishing eye contact, but the Active but Odds are all too disposed to it. In fact, they stare at others in an intolerable way. Fred Volkmar of Yale once told me that one of his patients liked to stare at girls he found attractive in a university dining room. Sometimes the young man would stare and stare until the campus police were called. Was he an Active but Odd? Active but Odds also like to hug other people with a suffocatingly inappropriate tightness. Combined with their tendency to stare hard at others, this habit causes endless problems for the Active but Odds. Their bewilderment knows no bounds.

Lastly, there are the Over-Formal, Stilted folks. These are the most able of all the four groups, and their symptoms do not appear until later in adolescence. "They try," says Wing, "very hard to behave well and cope by sticking rigidly to the rules." Yet these same rules are things they do not understand for one second. As their name implies, the Over-Formal, Stilteds are excessively polite and formal. When Over-Formal, Stilted boys want a date, they walk up to a complete stranger with excessive formality and ask for a kiss with the same manners a nineteenth-century Spanish aristocrat might have used in asking for a woman's hand. The offer is rarely accepted, to put it mildly; sometimes there are yet more unpleasant scenes involving the forces of law and order. We can take it as Aspergerishly axiomatic that the Over-Formal, Stilted, like the Aloof but Odd, are doomed to dating hell.

Wing's definitions are not severely precise. Perhaps most children share at least some of these characteristics, for they are often withdrawn, moody, otherworldly, solipsistic, fantastical, and opaque. The oddness of any given child should be sufferable. The question for the Active but Odds and the Passives is only at what point their active oddness and passivity become either dangerous to themselves or ruinous to a "normal life." But then again, of course, it is we and not they who decide what dangerous (or for that matter inappropriate) is. Can a Passive child be happy in his or her own way, liberated from the necessity of joining in our social small talk or its concomitant field of gestures? Can the Over-Formal, Stilted avoid the ostracism that comes with overt strangeness and difference?

For that matter, how well do we ourselves suffer eccentrics, allowing them their own small talk and gestures, a logic at right angles to our own? I vividly remember my Irish great-aunt who lived with our family throughout my childhood. A retired schoolteacher, highly correct and strict, she was a model of severe normality. One day, however, Auntie

Mary summoned me to her room at the top of the house, drew me sweetly to the window, and pointed down at the hedges surrounding our garden. "Do ye see them?" she whispered in my ear, pointing earnestly at the hedges and raising her whisper to an agitated hiss. "Do ye? Do ye see them?"

"See who?" I said, feeling a little alarmed.

"The *wee people*," she said, quickly making me promise not to say a word to my father about either our conversation or the *wee people*.

"Yes," I said, "I see them."

"Ay," she said solemnly, shaking her head. "They're out there every day with their shears. It's *they* who do the gardening, ye know." She took my hand and patted it; I was now a member of a secret club, the people who *could see the wee people*. I remember thinking at the time that I had now discovered something very dark, namely that our venerable matriarch Auntie Mary was mad as a hatter. But of course Auntie Mary was not mad. She merely saw the *wee people* in green waistcoats doing the garden where other people did not. I could not tolerate this at the time, but now I can. One cannot ask others to conform to a gray world in which the wee people do *not* prune the hedges of our gardens.

Of course, no one would suggest that Asperger children should not be helped to conform, because conformity is the price that they will have to pay to be able to hold down jobs, pay their rents, and put their names to marriage licenses. But the questions are still unavoidable, because they are the very same questions that are constantly being put to childhood itself. A child is never free to be himself or herself; he or she has to be socialized (that most folkish of all popular concepts). But the child's individual warp, the acute sense of apartness and loneliness that children have, is not a perversity that has to be flattened out. The child needs his inner perversities.

Thus a somber question arises. Will anyone, I wonder, have the audacity to leave that fruitful perversity untouched in the Asperger child—that very child who seems most to be in need of stern corrections, remedial severities, and an artificially constructed sense of normality?

★

Surfing the Internet, I had learned of a nine-year-old Asperger boy in California named Nicky Werner. Nicky had written a slender book of poems called *Thoughts*, which contained some strikingly haiku-like verses that were written out with the typographic form of a diamond. They were severe, compact, and unnerving, and above them Nicky rendered monochrome line drawings showing the simple element that was the subject of the poem. Like the poems, the drawings had a hard purity. One of them was about stars:

Star

big, bright

forming, dying, moving

makes me feel tiny

Sun.

These were quite unlike any poems I had seen from the pen of a seven-year-old, and I was intrigued. Nicky is the son of an Air Force recruiting officer who lives on an air base near Adelanto, California, in the Mojave Desert. I flew to Los Angeles with the intention of meeting him. But first I thought I would pay a visit to the head of the Los Angeles Asperger's support group, a man named Arthur Ringwalt. Arthur is well known in Asperger circles for his organizational abilities (a rare talent in those parts) and his strident views on the plight of autistics in the United States. I thought I might even ask him some questions about Nicky Werner, on the off-chance that he might know him. In any case, I arranged to drive to his apartment in Glendale and interview him. "Just let me know the time," a voice with what sounded to me like a dry Swedish accent said on the phone. "I'm an athlete, you know, and my tennis schedule is very strict."

As I was driving through the rush hour on Sepulveda Boulevard, I wondered why Arthur had a Swedish accent. Was he Swedish? His house stands on a quiet road called Elm, off Glenoaks Boulevard, where the old trolley lines rust between traffic aisles. Because it is Glendale, you can look up from the street to the San Gabriel Mountains, a narrow but towering pastoral vista. Arthur's condo is over a carport, and to the door came a wiry, tanned fellow in a purple Le Coq Sportif baseball cap and dark green tennis shorts. His eyes blinked and he waved me into a room of topsy-turvy clutter in which could be seen mountains of crates, boxes, and papers piled like the contents of a warehouse. No fan or air conditioner; we sat in the kitchenette, in the heat.

"So," said Arthur in his faint and rather staccato Swedish-sounding accent, "you made it. You're four minutes late. But I'm glad you came to the interview."

I saw now that he had a carefully typed questionnaire in front of him, the two sheets stapled together.

"I've made it brief," he went on, "so I hope you can bear with me. But I think you should read this first."

He handed me a single typed page headed by the words *Arthur Ringwalt, Brief Bio*. "Read it," he said.

The page began "I am single, male and 58 years of age. Gary B. Mesibov of the University of North Carolina diagnosed me with Asperger's Syndrome in 1999." A little further down, it went on: "I compensate for my disability. I am well organized, resourceful, and a good team player. Sports, particularly tennis, are an emotional release for me. . . . I am writing a pamphlet called 'Hire People with Autism: Autistics Are Good Workers'."

"So," he said. "What do you think?"

"It's succinct."

"That's what I was aiming for. I've also tried to be succinct in our interview notes. I have it all written out."

So Arthur was going to interview *me*. He cleared his throat.

"First question, Mr. Osborne. Do you have or have you ever had any medical qualifications whatsoever for writing about Asperger Syndrome?"

"None whatsoever."

"I see." He scribbled quickly. "Second question. What kind of book is it that you are writing?"

I said I didn't really know.

"I see." More scribbles. "Third question. What are your other qualifications for writing this book? Is it a scholarly work? Anecdotal?"

And so on. While this grilling went on I looked around the room. There was a picture on one wall of a line of camels parading past the walls of an ancient Chinese city, the whole scene tinted with the artificial colors of a photograph from the 1930s. Across the room, Alpine pastures; on yet another wall, sporting snapshots: baseball heroes, the Joe Louis–Max Schmeling fight. Then there were the boxes. I now saw that they were packed with scores of containers of weight trainers' protein powders. Lean Body, Met-Rx Engineered Nutrition, Hard Body, and bottles of Refreshe water. The coffee table groaned with muscle mags. I spied at once the formidable *Testosterone* as well as *Muscle Media*. There were scores of them packed in sheaves. On the kitchen table, on the other hand, stood a bottle of Martinelli wine with a box of After Eights shrouded in plastic gift wrap and a well-thumbed paperback edition of *The Idiot's Guide To Managing Your Time*. Just as I was settling in, Arthur wrapped up his interview and suddenly handed me his professional card. It bore a Corinthian column and the words:

Arthur Ringwalt
Classical Technical Writing

"But," he confessed sadly, "the market's dried up. These are tough times for us technical writers."

Arthur turned out not to be Swedish at all. His father was an American diplomat in China, and the Ringwalts appeared to be a fairly prominent family in Virginia. Arthur was raised in boarding schools, but his mother had already noticed that there was something not quite right about him. At two, he was taken to the Boston Children's Hospital where his mother received the cheerful recommendation that he be institutionalized for life. What was wrong with him? Nobody really knew. If he had autism, why was he able to function so normally at school, despite being constantly teased and despite his great love of repetitively swinging doors? Wisely, his mother disobeyed the men in white coats, who so obviously do not always know best. One can imagine at the time the severe reprobation that this act of medical disobedience must have provoked. But at least Arthur was free to continue in special schools, in his case one for brain-damaged children in Wisconsin. There was no evidence that he had suffered any kind of brain damage. But his Aspergerish traits were too baffling for him to be classified otherwise. And so he was left to immerse himself in his obsessions, according to which he occasionally thought he was Napoleon, became extremely knowledgeable about English country houses, explored the outer reaches of astronomy, continued to swing every available door in sight, and made himself something of an amateur chess master. ("I can be beaten when I'm not concentrating!") It was, in short, a miserable childhood.

"I didn't talk until I was six. But I do remember hearing Rachmaninoff for the first time. That was an event. Otherwise it was hard. Asperger childhoods were not much fun in those days."

Nevertheless, he had something of an academic career: the University of North Carolina, the University of Denver, a math degree. He went on to teach but couldn't control his classes. He switched to selling the *Encyclopedia Britannica*, but with little success. Prospective buyers of the august encyclopedia somehow didn't take to his strange accent, which is not a foreign accent at all but an Aspergerish lisp. "I didn't sell even one," he sighs. Supported by his trust fund, he drifted into "classical technical writing," whatever that is. And after he was diagnosed with Asperger's in 1999, he drifted on into the world of autistic advocacy, meanwhile avidly playing tennis in all his free time. In fact, athletics have more or less taken over most of his waking life.

"I live for it. Every minute. I am committed to putting on muscle."

He raised his skinny arms for a moment and his eyes fluttered, batted, as he echoed the word "muscle" a second time. He then explained that all he ate these days was muscle-building powder. Was it having any effect, I wondered? Arthur seemed to complement Alfred de Musset's assertion that a poet should not weigh more than ninety-nine pounds. Yet he went to the gym every other day and religiously drank his Carbo Max. "Relationships?" I asked.

"I think I've had three," he mused. "Let me think. One . . . two . . . yes, it was three. The best was an Israeli girl back in . . . nineteen-seventy-something. But she got sick of it. I had sex once with a girl in Denver, I think. I took a girl on a date once. Touched her breasts—big mistake." He sighed more heavily. "Big mistake."

Arthur was beginning to remind me of a character mentioned in *Goodfellas*: Jimmy Two Times, who says everything twice. He recalled that at school he had been expelled for writing a girl an admiring but impertinent postcard. The injustice of it still rankled him. Beyond that, the world of loving relationships was even more alien than it is to the rest of the population. His closest intimate was a guy called Myron who used to fly bombers during World War II. But all in all his friends, such as they are, seemed to be of less interest to Arthur than his diet. With rapid-fire stutters and sudden verbal salvoes, often quickly repeated, he described his expeditions to the local Costco supermarket and his fever-

ish purchases of 10-percent-fat chicken nuggets and the strawberry par-
faits at McDonald's. "As you see," he said mischievously, "I'm really
organized." From his nutritional diet, we proceeded to the medical kind.
He said he was on the antidepressant Luvox, which he took to subdue
his Obsessive Compulsive Disorder, and whose side effects fascinated
him. He proudly showed me the box, which duly warned against:
"Abnormal ejaculation, tooth decay, blurred vision, frequent urination,
nausea, gas and bloating, upper respiratory infection, vomiting."

"*And* you can't jack off," he added triumphantly.

What were his compulsions? For one thing, he rapped tables with
his pencils a great deal. He was obsessed with bank mergers. Then, of
course, there was tennis. Luvox, I wondered, for tennis addiction? But
there was also Greek and Latin: Arthur loves Greek and Latin.

"I love anything written by Italians. Dante, the Italian movies,
Fellini. I want to read the *Æneid*. That's my next project."

I asked him then if he had ever heard of Nicky Werner, out in the
desert. Arthur shook his head. How old was he? Nine? A tricky age. He
got up and began pacing around the room in a burst of agitation. It
wasn't easy for the small kids, he should know! How does one lead a
profitable life if one isn't shown every encouragement at any age?
Arthur's father had brushed him aside at that age because his oddities
had alarmed him so deeply that Ringwalt Senior couldn't face them
head-on.

"And what's this boy's name?"

"Nicky."

"And he's in Adelanto? That's in the middle of nowhere. Are you
going there now?"

I said that was the plan.

"Then," said Arthur, "you're going to need some liquid refreshment."

"That is very likely."

With that, the interview, if it was an interview, came to an end, and
we sauntered outside.

"It's an ugly street," Arthur remarked looking up and down placid
Elm. The street is actually like most streets in L.A., neither better nor
worse. Rows of little adobe-style villas with yuccas outside; I always
think of that sinister line in the Beatles song *Strawberry Fields*: "underneath the blue suburban sky." I said it was a pleasant enough street.

"Pleasant?" he retorted. "Who wants to live on a pleasant street?"

Who does? I thought.

"I sometimes wonder," Arthur went on before turning away, "why everything has to be ugly. Why does everything have to be ugly?"

I confessed that, really, I had no idea why everything had to be ugly.

"That's just it." He snorted. "Nobody has any idea why it has to be ugly. Sometimes I feel so sorry for neurotypicals. They're so screwed up."

Then he politely asked what disorder I suffered from—it was clearly inconceivable that a fellow citizen might *not* suffer from a disorder of some kind. So I said, "I've got a case of Suburban Ennui Disorder Not Otherwise Specified."

"So," he winked, "that would be SED-NOS. Never heard of that one. Is it rare?"

We stood for moment looking up at the burned crests of the San Gabriel Mountains. A sarcastic smile had taken hold of Arthur's mouth and remained there as if fixed by an unpleasant thought. Then he abruptly raised his hand in a kind of military salute and removed me from his field of vision.

"Bon voyage," he said. "And remember to take your liquids."

★

Adelanto lies two hours out of Los Angeles on Route 15. I had a strange feeling going into the desert: It was sharply nostalgic for me to come back to the Mojave for the first time in seven years, a place where I had occasionally worked as a reporter for a San Diego newspaper, trudging the state's back roads in miserable solitude while investigating long-forgotten stories on migrant farm workers, bingo casinos, and the date farms of, well, Date. Stories that probably no one ever read. Now as then, I wondered who on earth would want to live in such an unrelenting place. The desert is actually filled with surprisingly normal people who go about surprisingly normal business. I remember a piece I once did on a small town's fire department. Those huge bronzed men talked about putting out fires in the desert as if it were the most mundane task a human being could be asked to do. They had not a whit of romantic feeling for their surroundings. Why then, I asked them, did they live in such an inhospitable place? "Ah," they said, "ain't nowhere like the desert." The meaning of this reply, of course, lay in its tone: childishly forlorn.

The road to Victorville is like all the desert roads, adamant as a Roman highway in conquered wilds. The badlands are partitioned out between military shooting ranges and imposing correctional facilities, and in between lie the eternal arroyos with their Joshua trees, their ocotillos shaped like upended jellyfish looming over dirt roads. I suddenly remembered how lonely I had been on these same roads, trawling from dust bowl to dust bowl while learning my Mexican Spanish from a Belts paperback. It's a landscape that lays a fear upon you. I wondered if Nicky Werner saw its gentle, opulent side?

At this point I was confronted with yet another of my rather Aspergerish traits. It happens that I am insidiously drawn to a motel chain known as the Red Roof Inn. Anyone who has driven around the U.S. knows that every freeway, every town, every intersection offers the same constellation of chain motels whose signs light up the sky with a barrage of competing prices. I could not say how much a Hampton Inns, a Motel 6, or a Travelodge differs from a Red Roof Inn, but there is something about Red Roof Inns that makes me search them out with an almost frenzied insistence. If I stop for the night in the middle of an unknown state, it has to be at a Red Roof Inn—which I have been told, incidentally, is the worst possible choice among the cheaper chain motels. What, then, makes me always go to a Red Roof Inn? I have no great love of them, after all. Rationally, I can easily see that it is a very bad choice of motel indeed. In fact, I loathe Red Roof Inns. I have a veritable catalog of complaints that I bring against the entire chain, and that long and bitter experience enables me to substantiate. But when all is said and done, these objections are merely rational and have nothing whatsoever to do with the fact that I am always bound to choose a Red Roof Inn for my stopover. I will go miles out of my way if I see the Red Roof Inn sign, and as soon as I am on my way to one I am filled with a childish satisfaction and calm—a soothing fore-knowledge of the Red Roof Inn experience, which instantly allays all my anxieties about being on the road while tired.

But here in Adelanto, to my severe dismay, there was no Red Roof Inn in sight. This had me stumped at once. I felt myself succumbing instantly to a minor panic attack. I was forced to stay at the Day's Inn in Adelanto.

Along Route 395, rigs roared all night long lighting up the desert like ferocious express trains while I paced up and down. Finally I turned on the TV: Larry King. By one of those perverse alignments of which round-the-clock TV is marvelously capable, Larry King was interviewing the famous Dr. Phillip McGraw, alias "Dr. Phil." Dr. Phil is one of the self-help gurus who has forged a special relation with Oprah Winfrey, but here he was explaining his new book, *Self Matters: Making Your Life from the Inside Out*, to Larry King.

Bald, with a southern twang and continually twitching eyes, Dr. Phil is one of the maestros of Relationship Therapy and says "Let me tell you" every fifteen seconds.

"I've got more degrees than a thermometer," he was chuckling, "and let me tell you, that's not how I got so wise."

"What's your secret?" King asked, or words to that effect.

"Each one of us has what I call a personal truth. I ask people to write down their personal truth and make it fit it into the Ten Laws of Life. Larry, let me tell you, this book was a process book. I got into the process of it all—the process of happiness. We need to address our issues with our wives and reopen negotiations. . . ."

Like the mad preacher in *Night of the Hunter*, Dr. Phil droned on about processes, negotiations, positivity, stress and distress, love and hate, listening to oneself, having one's needs met, and yet more negotiations about issues. People called in from all over the country. *Hi Dr. Phil, do you have a formula for stress management? Dr. Phil, how can I get my needs met?* "This is Dr. Phil," King said, "and our lines are open right now."

Adelanto itself was just a scattering of wind-worn shacks. A Charlie's topless bar at the crossroads, a Burger King, boarded cactus nurseries. Overhead, an immense cat's cradle of power lines. The town's strip is Palmdale Avenue, and, as it crosses 395, the car lots and bail bond offices give way to vast housing tracts in which the venturesome white race lives as if sequestered in barracks—interlopers in a land so ill-suited to their skins, always floating around on its surface like fragile water boatmen, insects skimming over a treacherously dark lake. Roman citizens in the outposts of the Sahara two thousand years ago. One can imagine the same thing throughout history: an entire race injected into a land that instinctively spits them out. I think to myself, *How long will we be here?* It is in one of these severe residential developments that the Werners live. Their streets seem exposed to some terrible solar blackmail, with names like Delicious and Palm, and the houses struggle against malign, implacable elements. There is a purr of air conditioners that are never turned off.

The Werner house sits at the bottom of a somnambulant cul-de-sac. I felt a little like an astronaut descending from a shuttle to a friendly base on one of Jupiter's moons. What would one say to the stranded inhabitants? And there at the door was Trina, Nicky's mother, and her husband still in uniform after a day on the base. They are young, disciplined, intelligent—a model military family to my eye. He slim, quiet, reserved; she Mediterranean-looking, bodacious, extrovert. American military families are by nature rootless, and, unlike those of the majority of their compatriots, their perspectives on the world are often formed by the experience of living in many countries. The Werners said that they were pining for an eventual posting in Italy or Britain. Trina did not seem very comfortable in the desert or in Adelanto, though that may have been just my impression. She confided at once that she had loved living in Italy, where children are unconditionally adored.

"I wasn't very impressed," she said, "when I got back here. We don't love children in the same way Italians do."

There were three little boys inside, and Nicky was the oldest. As soon as I saw him, I thought to myself he was one of them, a classic Asperger boy. And indeed he was. Delicately unassuming, introspective, with huge blue eyes swimming behind professor spectacles. He shook my hand and told me gravely that he had no friends at school. I noticed at once that had a slight, well, foreign accent—or rather a faint trace of one.

"It's true," said Trina. "Up to the age of five he had a kind of Swedish accent. People thought he was Swedish."

At first, however, they thought he was deaf. Baby Nicky would never respond to his name being called, and although he had memorized words instantly when he was a tot, he barely spoke. Trina took him to a speech therapist. The therapist was nonplussed. Nicky couldn't speak easily, but he certainly knew all the names of cars, planets, and moons, the distances between the sun and every known planet and moon in the solar system, as well as the exact circumference of each. The Werners thought they had a little genius, but not an autistic one. After Nicky skipped kindergarten, he began to cause small, almost subtle disturbances at his school. He had to take things apart and see how they worked. He'd drop sticks into the water tank to see how the ripples played out. Again and again, an annoyed principal summoned him to the office. It's much the same story with all Asperger children: a wayward, solipsistic individual of high intelligence pitted against an inflexible system that cannot accommodate the odd, the not quite socialized, the defiant, or the mildly uncooperative, or even the slightly anomalous. It's a recipe for war, or at the very least for disruption and classroom disorder. The Werners had therefore decided to home-school him, and it seemed likely that Nicky was thriving with it, as most children do now. He already scored in the ninety-ninth percentile in his SAT 9 test.

We went into his bedroom. The desert comes right up to the housing development, and through his bedroom window I could see sand and cactus. The room was scrupulously neat. On its walls hung a small boy's obsession with outer space—an astronaut's dried ice-cream sandwich packet, "A View of Earth From Space"—along with a shelf of meticulously arranged mineral and shell specimens that Nicky proceeded to identify with professorial calm: iron pyrites, abalones, quartz, pine cones, opals, and lumps of mica. Each was carefully posed as if in a display at the Natural History Museum. We looked through his equally neat rows of books: the odd *Sherlock Holmes* and *Treasure Island*, but mostly fact books on the natural world, things like the *National Geographic Guide* to Seashore Life or Insects. Then Nicky brought out the printed version of *Thoughts*. As we leafed through the handsome drawings and dreamy haikus, I asked him if he liked living in the desert.

"I like the heat," he replied.

And what about the air force base? Did he know all about planes?

"Some of them. When I grow up I'll be an astronaut."

At the end of the book there was indeed a four-line bio which read: "When he grows up Nicholas wants to be an astronaut, get out of this place and live somewhere else." Was that how he felt? I asked. "Sure," he said. "Get out of this place—planet earth—and live somewhere else."

Then he asked me, "Do you like reading poems?"

I said that I did it all the time. It felt like a confession of a dubious habit, which nowadays it no doubt is.

"Short ones or long ones?"

I'd never thought about the issue quite that way. "Short, I suppose."

"Me too. Short is best." He fixed me in his drilling gaze, a complete inversion of the usual Asperger eye-avoidance. "What about Harry Potter?"

"Not really my thing."

"Harry Potter's almost as interesting as math."

In *Thoughts* I came across another poem I liked very much:

Earth

big, blue

supporting, reflecting, drying

makes me feel small

People.

After this came two others that dealt with war:

Country

huge, strong

defending, fighting, celebrating

makes me feel proud

USSR.

USSR

vast, gone

forming, ending, launching

makes me feel glad

Rocket.

Hans Asperger had claimed a penchant for metaphor to be a defining characteristic of his boys, and Nicky certainly seemed to have a bent, if not for outright metaphor then, for compact associations. His mind was both disturbingly hyperfactual and blithely associative. Undeniably there was something a little sinister in the war poems, as if he had hit on absurd emotions lying latent in the nation's collective airwaves, while he himself was merely "making sense" to himself as he wrote. On the other hand, there was nothing remote or stiff about him as he sat next to me on his bed showing me his drawings. The physical distance that Asperger children are said to always maintain between themselves and others was absent in him; in fact, he was affectionate and trusting. We peered through the window, and he told me the names of the plants sprouting in the dust. Although he sounded somewhat like a kind of miniature botanist—one of the more angelic creatures from Dr. Moreau's island (a scientist interbred with a child?), his manner was confiding. He exuded a definite pleasure in showing a stranger around his den. Afterward, Trina described his fetishistic attitude toward food. When Nicky's fish fingers are cut up on his plate, it's imperative that no piece should touch another piece. There will be exactly eight pieces cut to the same length, neither more nor less, and the fork he uses cannot be allowed to touch the surface of the table. He is also allergic to eggs, peanuts, and milk.

"This latter," said Trina, spread Rubenesquely on the front room sofa as the boys swirled around her, "was especially interesting to me. You see, I'm studying to be a lactation consultant!"

Stumped for a moment, I asked her if she thought diet played a role in Asperger Syndrome.

"Genes," she said flatly. "After all, I *breast*fed Nicky."

She explained that her husband's half-brother was exceedingly strange. The father was a preacher and the mother a funeral arranger. Together with the half-brother they formed a kind of mobile evangelical funeral service. Religious fanaticism, as far as Trina was concerned, had made the half-brother even stranger than he already was. Did he too have Asperger's? No one had ever asked.

As I was leaving, Nicky told me that he had an inordinate fear of germs.

"They're everywhere! Staphylococcus."

"What about in the milk?"

He smiled shyly. "Especially in the milk."

I asked him if he wanted to come outside, but it was 102 degrees, and he shook his head. The family stayed indoors through the summer.

He asked me where I was going next, and I said San Jose. There was another Asperger boy there named AJ who was a couple of years younger than himself. At this, Nicky's interest perked up.

"Is he a military boy?"

"No, he lives with his grandmother."

"Is he normal?"

"I have no idea. I'll have to see."

"I wonder," Nicky said, "if he's interested in Jupiter?"

I said I would be sure to ask him some questions about Jupiter.

"Do you think he knows how many moons Jupiter has?"

I was already walking down the path to the car, cringing in the heat, while Nicky waved from the front door holding an iron pyrite in one hand.

"He might," I called back.

But Nicky was very serious, standing on tiptoe.

"Well, you ask him. Ask him how many moons Jupiter has. *I* know how many moons Jupiter has. I'll bet he does know. You ask him!"

★

San Jose is a city of handsome banks and relaxed ghettos, its streets shaded and vaguely Mediterranean, though devoid of any café tables. The San Jose Day's Inn reminded me very precisely of the St. Louis Airport Hilton. Every corridor mirrored every other corridor, while the rooms reveled in the same mass-production feel: sterilized water cups and fake bronze lamp handles, acrylic pictures of English country roads and iris-motif bed covers, stiff pleated curtains, courtesy soap packs, and racks of wire hangers. *The moons of Jupiter*, I kept thinking as I watched that night's *Iron Chef*. So far I knew nothing about this second little boy, AJ, except that he was interested in vacuum cleaners and not the moons of Jupiter. *Do the Mindblind*, I wondered, *have an affinity with each other regardless of their environment, attracted perhaps to the very idea of an obsession?* But this raises an unanswered question: What exactly *is* an obsession?

The notion of an unusual perseveration is becoming less and less abnormal as millions of people are diagnosed with OCD. The moons of Jupiter, vacuum cleaners, *Iron Chef*: we are all implicated. In fact, I watched with rapture as Grand Master Mitsua Harada of the Brighton Hotel in Kyoto came out to face Iron Chef Chen Kinechi, the king of Chinese cuisine. A superb gloom hung over the proceedings. Facing the camera, the demonic compere Takeshi Kaga picked up a yellow pepper, grinned sarcastically as if about to bite off the head of a small child, and sank his teeth into the pepper. I cannot say why this Nipponic kitsch enthralls me. I suppose it's the idea of the horrible, difficult raw material, which is always unveiled at the last moment, and which imposes upon the masters half an hour of exquisite torture. This time it was a battle with Spanish mackerel. A steaming pile of giant Spanish mackerel rose into the kitchen, and the masters began to sweat. The commentators, including the extraordinarily beautiful Chiruza Azuma, became heated. "Why, he's wrapping radishes around chopsticks to make them curly!" "It looks like a Kudzu starch sauce served with

Spanish mackerel canapés!" "Unusually avant-garde but rigorously Kyotan!"

Harada's victory gave a sudden rush of pleasure, and I danced around the room. It must, I thought, have been the whole grilled Spanish mackerel with hot bean sauce that had clinched it. But hadn't Chen Kineshi, the mightiest of the Iron Chefs in his yellow satin hat, gone undefeated for nine months? Grand Master Harada's victory, therefore, had a special historical significance. What had the bean paste grilled on the spine of the whole Spanish mackerel tasted like, or the Spanish mackerel porridge-style? What of the grilled Spanish mackerel laid on slices of lemon, or the Spanish mackerel skin grilled and served Peking Duck style?

And here is the Aspergerish rub of my *Iron Chef* perseveration. All the Iron Chef battles are organized around a single foodstuff. They are manic variations on a single theme, which is exactly how we don't eat real food. It is, however, a well-known characteristic of Asperger people that they will eat only one foodstuff, sometimes for years on end. Apart from Glenn Gould's infamous arrowroot biscuits, I remember a case history related to me by Yale researcher Fred Volkmar during a telephone conversation. One of his patients had invited a woman to his home after she had reluctantly accepted to have a date with him. Volkmar asked the Asperger man later how this dramatic event had gone—success or failure? The man was baffled. He had made an elaborate dinner for her, but as he had presented with some pride his elaborate third course the woman had burst into enraged tears, uttered an expletive, and fled the premises. "What," Volkmar asked tentatively, "had the chef served her?" "Well," the man answered, "I only eat sardines. The sardine is the perfect nutritional food. The ratio of fat to protein is optimal. So I served three courses of sardines. Sardines with baked tomatoes, sardines with melted cheese, sardines . . ." Volkmar added that the man couldn't understand why people would eat anything but sardines—cereals, for example, or pastas. They weren't as nutritionally efficient as sardines; therefore, eating them made no sense. Besides, he was used to eating sardines, and sardines were all he was going to eat. To eat anything other than sardines would be to change the rules, and changing the rules was a trauma.

The Aspie's mania about food (which is never enjoyed as a carnal pleasure or even as an art form) has an interesting mirror image in the wider culture. For the latter, too, is literalistic when it comes to food, generating hundreds of madly radical diet regimes and food-related hysterias. There is something distinctly Aspergerish about the way Americans insist that all foodstuffs carry a precisely analyzed list of proteins, carbohydrates, fats, and vitamins on their wrappings. Arthur Ringwalt, after all, was not so very different from followers of cult diets such as Calorie Restriction—a severe diet intended to prolong life expectancy—with their maniacal pedantry about the "calorific values" of every raw carrot, raisin, and alfalfa bunch they eat. Some obsessional dieters even have special kitchens built into their basements to create their outlandishly severe meals. Yet such attitudes are considered largely normal.

But the Aspie's refusal to eat anything other than one thing also has a kind of crazy purity in which the Iron Chefs also indulge. This obsession lies with wanting to exhaust all the possibilities of a single foodstuff. It is rather like Arthur's interest in bodybuilding protein powder drinks or a San Jose boy's tenacious interest in vacuum cleaners. Pleasure is a difficult concept for Asperger people to grasp because it involves a notion of play, of *appearances*. In a world interpreted literally, but without rules from which they can depart with dashing spontaneity, they are nailed to their fixations like dead moths to a board. But then, I must humbly add, so are we all.

Trish Canepa lives in the Pepper Tree mobile home park off Monterey Avenue in Santa Clara County. Under the imperial palms, the units seem less gritty than they actually are. For a moment you wonder if you are in some Mafia motel in which there must be a casino and a slot machine hall somewhere. Trish is only fifty; she ran her own real estate company in San Jose before her daughter Dina spiraled into prostitution and drug addiction and lost control of her four small children. AJ was born in the Sunrise Hospital in Las Vegas with 1000 ml of amphetamines in his blood, as well as 300 ml of opiates, barbiturates, cocaine, and an equivalent amount of benzodiazepines.

Hyperactive, nervy, he met me at the door with his teddy bear, his gaze slightly skewed—crossed, even. He told me at once that if I gave him two dollars he'd give me two dimes back. Then he asked me if he could use my keypad to set off the alarm in my car. Three seconds later, he was pulling me into the yard to show me the dirt patch he liked to dig up with a plastic shovel. I felt buffeted back and forth, as one does when playing with a muscular puppy. He said, "I like to go to sleep with stones in my hands. If I don't have stones in my hands I can't get to sleep."

"It's true, he has to sleep with stones in his hands." Trish was matter-of-fact. We sat in the front room and AJ whirled around the unit, running from one end of it to the other on bare feet. AJ, she added, was first diagnosed with ADD before being identified as having Asperger's by the Children's Health Council at Stanford University. In kindergarten, at age five, he broke the nose of one of the mothers and was expelled; on another occasion, his school called the police to forcibly remove him from under a bus seat. "He throws terrible tantrums. But the vacuum cleaners calm him down!"

Trish is a reminder of how gutsy solitary grandmothers so often hold together a disintegrating social fabric. She married a scrap-metal dealer in her teens and had Dina when she was eighteen. At two, Dina was already out of control; at ten, they took her to a psychiatrist. "She was a sociopath, always stealing, lying. When she was sixteen, she was doing drugs heavily. She was just like her father, actually." Trish moved out when Dina was twelve and remarried. Back in Tahoe where the father had kept the children, Dina was soon convicted for aggravated assault and burglary. Thereafter she quickly progressed to prostitution. In and out of prison, she had four children with a variety of men. AJ's father was one of them, but Trish said that he remains an unknown.

"For all we know, it was one of her clients. It could have been an Oklahoma senator."

With no parents, AJ was at sea. His little sister is also disturbed, diagnosed now with OCD. Trish's common sense tells her that all these diagnoses of medical conditions are only tapping into the disturbances of a profoundly abnormal environment. But the disturbances are nevertheless as real as the rocks that AJ holds in his hands when he goes to sleep. ADD, OCD, Asperger's: the litany of childhood disorders provides the harassed parent or guardian with a kind of map with which she can orient herself as much as the child. Yet at the same time there is no easy escape from these same afflictions, no commonsensical road out.

After AJ's various diagnoses, the inevitable drug regime began. The psychiatrists prescribed Dexedrine, which AJ now takes daily with a 50 mg dose of Luvox and a 1.5 mg dose of Risperdone. They warned Trish that he might lose two inches of his normal height with lifelong use of

Dexedrine. Terrified, she began to have doubts. Ritalin was tried, but with disastrous results. On the other hand, she could see that the Dexedrine had helped him expand his normally ephemeral attention span so that he could actually finish a page of his coloring book. When the Dexedrine was cut, AJ spiraled out of control again. This presented Trish with a dilemma. Dexedrine may have helped AJ stabilize, but looking back at photographs of him taken before the regime began she noticed at once that then he was always smiling. Now, he doesn't smile at all. "The drugs have taken all the joy out of him. Something has been sucked out of him."

AJ was certainly agitated and stressful. He blinked impatiently, flitting from activity to activity without any noticeable change of emotional inflection. The conversation inevitably came around to vacuum cleaners. AJ loved the promotional video that came with the new Phantom model and watched it over and over, while rocking himself back and forth. To discipline his moods, Trish would occasionally snap at him that he wouldn't be able to touch the new vacuum, and a sullen look of castigated impotence would suddenly come over his face. The threat clearly worked. As we sat in the front room, he crushed a few soft rocks from the yard into the carpet, despite Trish's warnings, and then swiftly disappeared to fetch the Epic Series 3500 model from his bedroom. He gets a vacuum every year for his birthday. Taking up a dark red toy magnifying glass, he then crawled around the carpet with the glass pressed closely to the fibers. He explained:

"There are bad devils in the carpet that have to be vacuumed. It's gonna make mom feel much better." He calls Trish mom. "It'll capture every little one of them. You need the magnifying glass to see the devils and the baby dirt. That's what I love about vacuums. They go everywhere, you can take them anywhere, and you can pick up dirt even under tables."

Turning the vacuum on, he then began furiously running it over the carpet in all directions.

"Can I vacuum in the hallway?" he pleaded with Trish. Given permission, he was ecstatic.

"See how fast I picked them up with the bristles?" He then came sidling up to me and whispered very quietly in my ear, "If I use the vacuum I can get it out quicker!"

It is difficult to draw conclusions about that twilight zone where environment and genetic makeup collide. I thought about the differences between Asa, Nicky, and AJ—three boys diagnosed with the same specific neurological disorder, which should have made them much more similar than they actually seemed to be. Each one, of course, mirrored a specific environment, for good or ill. What, though, would be the kernel of their future personality? Is there, one has to ask, such a thing as an Asperger identity, just as there is presumably a schizophrenic one?

The child is always malleable and adaptable, and AJ is now in a remedial school for autistic children, where he'll probably thrive eventually, even if he ends up being something of a social misfit. As he walked me out to the car, he continued trying to cut little deals with me. Would I give him this if he gave me that? A quarter in exchange for a yard stone? Eventually I relented and gave him the quarter. He ran around the car squealing, "Yes! Yes!", then activated the alarm from the keypad. For a moment, minor chaos erupted in a quiet lane of the Pepper Tree mobile home park, and one small boy was in ecstasy. But Trish had to call down order.

"If you do that again," she warned, "I won't let you play with the new Phantom when it arrives next week!"

Immediate silence.

"You're my friend," he whispered in my ear. "When you come next time, I'll let you touch the Phantom."

I said that would be swell.

"You have to see the Phantom. It's the most amazing vacuum cleaner in the world."

"I'll bet it is," I said.

"You wait and see."

I drove off, crossing an abandoned one-track railroad line obscured by the smoke of grass fires. A few shabby men in undershirts hung around their rusted Le Sabres playing cards. For a moment, I had a compulsive desire to go out onto the mall nearby on Monterey and buy AJ a brand-new vacuum cleaner. For AJ, I mused rather preposterously, the humble vacuum cleaner was part of a sort of "song of innocence," which William Blake would have understood at once.

★

Afterward, I drove along Route 1. I had wanted to visit Henry Miller's house near Big Sur, or at least the queer little cliff-top museum that now is his temple. But after having lost my way (an almost impossible feat since there is only Route 1 to navigate), I decided to stop somewhere between Big Sur and Pismo Beach, completely lost in the dark and suddenly aware that my ability to remember maps was one that only applied to certain countries and to certain states, but not to Route 1.

I knew only that I was a couple of miles from the Hearst Castle, which I had always wanted to visit. Yet it was July 4, and fireworks lit up the sky above Pismo Beach; anyone not afflicted with Nocturnal Disorientation Syndrome (NDS) would have found the sign for Big Sur easily enough. There was only one thing to do, namely look for a Red Roof Inn. But here at the edge of the Pacific there was none to be found.

In any case, I did come across a silent place wedged between the road and the cliffs called the Jade Motel. Its lights flickered, huge signs crying *No Smoking* hung amid colonial clocks. *Psycho*, I thought. The Bates Motel! An old woman in an incredible plastic headscarf took me upstairs. "It's a shame you're here," she said. "You should be enjoying yourself in Pismo Beach." I rolled into bed in a numb fit of disillusionment when I had discovered at once that the room's TV set was no longer functional. This meant no *Iron Chef* and no Weather Channel for twenty-four hours. Furious, I was unable to sleep.

On the walls of the room hung some forbidding paintings by a man named Carl Hasch; Turneresque imitations of rustic English scenes I should have known from life, but didn't: fords, stiles, shaggy horses, barley fields with ruddy swains in frocks. For some reason, they too bothered me, as if I had seen them somewhere I could now no longer recall.

Outside, the Pacific ground away against the back gardens of a row of suburban chalets. Alongside the chalets stood a lofty row of lampposts that cast a leprous light into the back part of the room, and they were keeping me awake as well. Finally, I got up and went outside down to the cliffs where the lamps were providing a buffer against the glittering lights of Pismo Beach. It was now two in the morning.

I went around each lamppost counterclockwise, making my way as far as the edge of the cliffs and looked down into mounds of driftwood and kelp. I had no idea what I was doing there. But, as I have explained,

I have a thing for lampposts: I cannot resist them, they draw me to themselves like great stork-like sirens. As soon as I am near a lamppost, I feel a resurgence of some childhood impulse to walk around it counterclockwise, to taunt, antagonize, and "ensnare" it. In short, I become abnormal around lampposts. This could be otherwise put, however: for I could just as easily say that lampposts have the same effect upon me as certain (to us) invisible navigational signs have upon homing pigeons. They are markers—signs that cannot be explained. And they trigger deep reactions that also cannot be explained.

When I came back to the Jade Motel—yes, exhilarated not by the moon shining on the kelp beds but by a simple colonnade of lampposts—the woman in the plastic hood popped out her head from the night office and scowled. She had seen everything, of course. Clearly irritated by my display of abnormality, she cocked her eyes toward the beach and, by implication, the lampposts, and said, "It's dangerous down there, you know. I wouldn't go down there, if I were you." She elongated her words as if, being a retarded foreigner of some kind, I could barely understand even the simplest word. "It's not meant for recreation. It's private property."

The next morning in the bookstore at Big Sur I picked up Henry Miller's study of Arthur Rimbaud, *The Time of the Assassins*, and retired to the cliff-top Nepenthe Café across the road. This was not quite fortuitous. I had been thinking all the way what kinds of adults these Asperger boys would grow up to be. Wasn't this the same question that you ask naturally about that quintessential child genius Rimbaud? If the kingdom of childhood has become threatened to the point of becoming extinct, how much more true is this of that other, equally mysterious category: genius? As we have seen, genius is also inextricably bound up with the symbolism of Asperger Syndrome. Some have even called it a "genius syndrome." The Asperger person is held to be a prodigy, a savant, a nucleus of extraordinary abilities neurotypicals can only marvel at. "Civilization," Temple Grandin has remarked, "would pay a terrible price if the genes which cause autism and Asperger's Syndrome were eradicated."

This may not tally with the realities of most struggling Asperger adults, or with the struggling parents of AS boys, but it is certainly the tack that Sacks took with Temple Grandin herself, and it is without question very popular among Asperger people themselves. To be exceptional, they argue, is a curse. But the flip side of the curse of having Asperger's is to be exceptional. How does the prodigy live in a normal society? Miller, being himself a kind of *monstre sacré*, asks the same question about Rimbaud.

Considered by any objective standard, the life of Rimbaud was one of catastrophic dysfunction, to put it mildly. It consisted precisely of a calculated warfare between himself and his society, a warfare that went far beyond the usual bohemian antics. For Miller, this could not have been otherwise. Technocratic civilization is so decadent that any exceptional being rebels against it as a matter of course:

> . . . With what ghoulish glee, when it comes to shovel him under, do we focus attention upon the "maladaptation" of the lone individual, the only true rebel in a rotten society! Yet it is these very same figures who give significance to that abused term "maladaptation."

Reading this slim book I remembered what it was that had attracted me to Asperger people in the first place. It is the fact that they are lost, that they are unknown to themselves. They are all a bit like Rimbaud running illegal guns in the deserts of Abysinia, estranged from country, family, lovers, friends, and gainful employment. That they are not lost for the same reasons as Rimbaud (though, who knows?) is of no importance.

Inevitably, one also asks whether this quality of being overwhelmed by an inner compulsion, of being a victim of oneself, is what we popularly mean by the word *genius*. It is, indeed, the very reason that we do not believe in geniuses anymore. For such a person is the reverse of ourselves, and we cannot bear anyone to be the reverse of ourselves, let alone superior to us.

Reading about Rimbaud—the teenage prodigy, the maladaptive loner, the spewer of metaphor—I could not help thinking about that other child prodigy and maladaptive loner, Glenn Gould. The question of genius is more disturbing than we think; it is, in any case, the next theme we must enter in our pursuit of the enigmas of Asperger Syndrome. But, as I whiled away an afternoon in the hills of Big Sur—like those of a gnarled, disappeared Greece before *moussaka* and topless night clubs were invented—sitting on the tourist terrace of the

Nepenthe, which, sadly, Miller would no longer recognize, I came across a strange brief paragraph that made me think of the collapsing dichotomy of child and adult that so hauntingly seemed visited on Asperger's orphaned children, who seem so bizarrely to resemble old men: "To the anabasis of youth, [Rimbaud] opposed the katabasis of senility. There was no in-between realm—except the false maturity of the civilized man. . . ."

CHAPTER 3

THE LAST PURITAN

People are about as important to me as food. As I grow older, I find
more and more that I can do without them.
—Glenn Gould

On a peppery and silver spring night, I was driving from New York to
Ottawa in a rented car with *Goldberg Variations* playing on the CD
player. I was thinking about Glenn Gould as I drove along Route 417,
which shadows the American border and is the same road that Gould
himself used to navigate in an unmanageable Cadillac on his way to
numerous concerts and recording sessions in New York. I was thinking
two things. Firstly, that Gould was the least classical of all classical
musicians; and secondly, that Gould was a hopeless driver and always
had accidents on this very same road—he could barely control the large,
luxurious cars he so vainly adored. But as I played his 1955 recording of
Goldberg Variations over and over, I thought of this moonlit flatland
dotted with barns and reflective marshes as Gould's very own manic-
depressive landscape—the Idea of North, as he called it. The incredible
precision of this famous recording (which made the twenty-two-year-
old Gould famous overnight), its contrapuntal clarity, goes well with
the subdued and slightly menacing emptiness of Route 417: nothing
distracts from the music's mathematically romantic concentrations.

I arrived in Ottawa a little after midnight. It's a spacious city of
Gothic papier-mâché, English-colonial revival. It looks like a glum

second-league English university town. Green and gold crenellations, spire lights, frosty crockets, all built like a theater set around the phlegmatic Saint Lawrence. You can easily imagine the Jesuits in their black robes pacing up and down the pebbled shores. I liked it at once because it is so different from the cluttered claustrophobia and megalomania of New York. Gould had never lived in Ottawa and would never have lived in Ottawa because it was too small, too provincial for him. But it was nevertheless extraordinarily Canadian, and Gould loved Canada. He could never have lived anywhere else. He loved Canada for its coldness, its arctic light, its spaciousness, its sense of primordial peace. Canada is more Indian as well as more English than the United States, but its Englishness is deliciously mournful, off-center, and skewed. Gould once wrote a song for a singer he was in love with, called *Das Kind der Rosmarie*, or "Rosemary's Child." His instructions for its proper performance, in German, were *"mit großem Gefühl und seelischer kanadischer Ruhe"*: with great emotion and soulful Canadian repose.

After spending an hour on motor patrol, marveling at the origami cityscape, I found my hotel near the Natural History Museum. There appeared to be no one there. But as I was checking in, in the gold-plated lobby, I found that a small and delicately greased man in a summer suit was standing next to me with a huge grin on his face. We said "Good evening," and he carried on grinning. Was he, I wondered, grinning at me, or just at things in general? It was difficult to say. The clerk asked me what I was doing in Ottawa. I said that I was here to look over the effects of the great Canadian pianist Glenn Gould, who was rumored to have AS.

"AS?" the clerk said knowingly. "Yes, I heard that too—AS."

"It's a rumor that just might be true."

"One of *those*, eh? I'll bet it was true. I'll bet it was."

"Excuse me," the greased grinner now said, looking very pleased with himself. "Did you say AS?"

"I did. Many people think he was AS."

"You don't say? May I give you my card?" He pulled one out and impressed it. (He was a farm feed salesman on the rove in backwoods Ontario.) "I have a special interest in AS. I have it myself. Oh, I know you won't believe me, but I do. It's a coincidence."

I looked into his merry, epileptic green eyes and did my usual double take. A lonely salesman at one in the morning looking for an equally lonely traveler? Was the friendly overture sexual, merely conversational, or just an antidote to mind-numbing boredom? I was wary. I also had the strange intimation that we might just be the only three people awake in Ottawa at that moment. Was this an onerous responsibility or not? I decided to let fey curiosity have her way. After all, if Mr. Pring (shall we give him the name of a noted autism researcher in far-away England?) had AS, then he was fair game for my purposes.

"Well," said Mr. Pring, still impaling me on that never-tiring rictus, which was looking ever more Frankensteinish with every passing minute. "Since I arrived at the same time as you, may I suggest a late-night drink? You must have a lot to ask a man like *me*."

We sauntered through the chilly spring night to a boulevard lit with lines of glass orbs, which must surely have been built by the French. Off to one side were cul-de-sacs lined with fast-closing restaurants and bars. Getting a table, I ordered a bottle of Sancerre and started wolfing it down. "I know the feeling," said affable Pring, "thirsty, thirsty. Of course, if I drink too much my symptoms start playing up." So there was a useful topic of conversation. What symptoms of AS did he have, I asked.

Pring seemed inordinately happy, amused, tickled. He rolled back in his chair and picked at two muskrat eyebrows.

"Oh, let me see, let me see. I have a lot of symptoms. I'm crawling with them. Let me see. Stereotypical behavior? I have that, oh yes. Ataxia of the gait? Oh yes, just ask my wife! Hypermotoric behavior, short attention span? Definitely. Behavioral uniqueness, bad verbal communication skills? The clinchers." He beamed, nodding, all teeth. "And then I'm happy all the time. I can't stop smiling. That's the final proof as far my doctor's concerned. And as far as I'm concerned, too, if I may say."

"I wasn't aware," I said, "that being happy was a sign of having AS."

"Oh yes, yes. Involuntary happiness. It's symptom *numero uno*. We AS people suffer these happiness seizures. We smile and smile; it's like lockjaw. Terrible. And there's no drug you can take to make you stop smiling. They call us the Puppet People, you know."

"They do?"

"You didn't know?"

"But in Asperger Syndrome—"

And here Mr. Pring's amusement level soared, and he began to howl gleefully, covering his mouth with a hand and wagging a finger back and forth. Between gasps, he ordered a plate of mussels.

"Sorry, but I just adore mussels. Asperger, what? I'm Angelman Syndrome. AS. And I assure you that they *do* call us the Puppet People. I should know."

"Angelman Syndrome?"

"Angelman. An English doctor, you know. Are you from there as well? They've done a lot of groundbreaking work on AS over there. You should be proud."

The Angelman Syndrome Foundation—for there is, indeed, such an organization—explains that in 1965, Harry Angelman, an English pediatrician, first described three children with the telltale characteristics. All had a stiff, jerky gait, were very slow to develop speech, laughed excessively and at inappropriate moments, and were prone to seizures and a host of other maladies. Other cases were eventually published in the medical literature, but the condition was considered to be extremely rare, and many physicians doubted its existence. Dr. Angelman relates the following:

> It was purely by chance that nearly thirty years ago three handicapped children were admitted at various times to my children's ward in England. They had a variety of disabilities and although at first sight they seemed to be suffering from different conditions I felt that there was a common cause for their illness. The diagnosis was purely a clinical one because in spite of technical investigations, which today are more refined, I was unable to establish scientific proof that the three children all had the same handicap. In view of this I hesitated to write about them in the medical journals. However, when on holiday in Italy I happened to see an oil painting in the Castelvecchio museum in Verona called . . . "A Boy with a Puppet." The boy's laughing face and the fact that my patients exhibited jerky movements gave me the idea of writing an article about the three children with a title of Puppet Children. It was not a name that pleased all parents but it served as a means of combining the three little patients into a single group. Later the name was changed to Angelman Syndrome. This article was published in 1965 and after some initial interest lay almost forgotten until the early eighties.

So there we have it: A Syndrome Is Born. Mr. Ping was indeed correct. People can be diagnosed with Angelman Syndrome if, in addition to having the usual social difficulties, they also appear to be too happy. This latter awkward pathology would, in any case, single anyone out as being afflicted by a disorder of the nervous system or even of the soul. But I am being facetious, and the syndrome is no laughing matter. Mr. Pring did not have jerky, awkward movements, but after a while I did indeed have the impression that something about him was a little "off." His laughter did indeed imitate the clinical descriptions provided 35 years ago by Dr. Angelman. Was it because, according to the information he himself had supplied, I had already decided that he was ill? Or was it because (an innocent possibility) he was an impressible, gay, good-tempered type who merely liked to smile and laugh? All in all, I couldn't help feeling that the Great Asperger Fairy in the Sky had played upon me a delicious and indeed rather typical trick, one that was in harmony with the law of absurdity that governs us all.

"And what on earth," said the ever-grinning Pring, "is Asperger Syndrome?"

"That," I replied, "is a long story."

Pring (laughing): "So? We've got all these mussels to eat. Was Asperger related to Angelman? No? But surely, Professor Osborne, they must have known each other, or at least known each other's work? Surely? No? Well, I never. I'm astonished. I must say, Dr. Osborne, that I find that entirely hard to believe."

★

The following morning I got up early and walked to the National Library of Canada. It's the typical postmodern affair—blocks, cubes, concrete: our nonlovely, nonlaughing nonarchitecture. I have a sometime office at the Bobst Library on Washington Square in New York, and the style is similar. You, the library-goer, are expected to feel much like a small functionary arriving at the centralized book depot of some awkward totalitarian state. At Bobst, you are confronted with metal security stiles through which the masses churn after swiping their security passes through magnetic slots. It is all very warm, as befits a temple of literature and knowledge to which you are making your small pilgrimage. At the National Library, I had to fill out the usual security forms, read a page of regulations, and fill out a questionnaire under the watchful eye of a guard who made sure I scanned every sentence, sign a one-day contract, and have a laminated badge pinned to my lapel. This was all in case I might be a wild book-hating terrorist hell-bent on slipping a Semtex bomb behind the early French monastic manuscripts shelf. Cleared by security, then, I wandered up to the fourth floor where the Music Division lies. There I was met by its director, a spry gray-haired scholar named Dr. Tim Maloney, who also had a laminated badge affixed to his lapel. In addition, he sported a splendid, if not extravagant, crimson tie with a boldly emblazoned head of Mozart on it, along with a few staves of music, presumably the immortal Master's.

"So," he grinned, "you got through security?"

Maloney is an obsessional but erudite Gould scholar who is currently writing a book that he thinks will positively prove that Gould suffered from Asperger Syndrome. For years, he has been amassing evidence of the Gould–Asperger connection in the face of fierce hostility and skepticism emanating from the Gould fan movement, which tends to like its Genius eccentric but neurotypically normal, if only barely. The Asperger's thesis, in other words, is not popular in Gould-land. But, no matter. Tim is on a crusade, and he is going to make his point.

"I'd say the evidence is pretty overwhelming at this point," he said, as we passed through a muffled exhibition area with large glass cases. "People can only go on denying it for so long."

In one of the cases, I immediately noticed Gould's famous Munchkin chair. This is a tiny collapsible card chair only fourteen inches off the ground, which Gould's father had built especially for him, and which Gould used all his life for playing. A legendary object, it looks, like most legendary objects, somewhat diminished and sad in real life. Beaten up and scuffed, its adjustable brackets and worn slats look more like a contraption for killing chickens than the beloved stool of one of twentieth century's most illustrious pianists. Gould insisted on using it, because it placed his arms at an unusually low height above the keyboard (the usual seat is nineteen or twenty inches high), and thus enabled him to assume that crouched, spidery posture of tactile intimacy he needed in order to play in exactly the way he wanted.

"Amazing, isn't it?" Tim cocked an eye at the chair. "Right there in that chair you can see something odd."

Tim is an excellent clarinetist and once performed under Gould when he conducted a Wagner chamber piece in 1982, shortly before his death. It did not go swimmingly. "He had a natural gesticulativeness," Tim recalled. "But he was untrained for conducting. He didn't use a stick—just these flowing, sweeping hands which were always vague and unfocused. It didn't convey much information. No staccato. Well, we just couldn't follow his beat, we couldn't synchronize." Gould's conducting, he went on, was very much in the nebulous style of Herbert von Karajan, whom he greatly admired. "But in this case, alas, it seemed to leave us alone as well as himself. It was very, well, autistic!"

Tim knew autistic children and was familiar with their characteristic behaviors. The sight of Gould rocking repetitively at his piano and moaning to himself as he played reminded him of these same children. Gould's notorious decision to stop playing before live audiences in the 1960s seems to Tim to be a sign that his Asperger's was getting the better of him. The solitary, one-on-one world of the recording studio suited Gould better because it removed any need to perform socially: always a daunting prospect for an Asperger person. Similarly, Tim went on, Gould's relations with women were more a matter of lacunae than substance. For instance, Gould only wrote one love letter to a woman, or at least only one letter that has survived. Tim thinks this cryptic text is addressed to a woman named Dell, the wife of a famous conductor in New York, with whom Gould was secretly involved. What is remarkable about it is that Gould doesn't actually address the woman herself: he uses an aloof, arcane third person.

> *You know*
> I am deeply in love with a certain beaut. girl. I asked her to marry me but she turned me down but I still love her more than anything in the world and every min. I can spend with her is pure heaven; but I don't want to be a bore and if I could only get her to tell me when I could see her, it would help. She has a standing invit. to let me take her anywhere she'd like to go any time but it seems to me she never has time for me. Please if you see her, ask her to let me know when I can see her and when I can. . . .

Tim saw this as clear evidence of an Asperger mind. And it's true that I have since come across many Asperger men who seem to write and think about women in this way. That is, with a kind of lost, disconnected hopelessness that betrays a mechanical inability to enter into the fray of sexual combat, with its unpredictable bolts of wayward electricity, its swift turns of mood and fortune, its undercurrents of masochism, mistrust, and fear.

Still, Gould's oddly stiff and even clichéd letter is tense with emotion. It may seem autistically remote, or else—if one chooses to see it that way—awkwardly but delicately restrained, like a courtly troubadour complaint. A missive certainly doomed to cut little ice with a contemporary American woman, one would assume. But it is, in any case, infinitely superior to the presumably neurotypical comments of Gould's accountant, Patrick Sullivan, who noted crudely that "he shacked up with a broad for about a year, some conductor's wife. I know, I saw the expenses."

I said to Tim, as we sat in his office, that I found Gould's tormented romanticism infinitely preferable to the dryly deconstructed sexuality now *de rigueur* in the advanced nations, with its dead but coercive language of mutual therapy and contractual dread. "At least," I said, "Gould doesn't 'acknowledge his issues,' or treat sexual love as a surrogate for economic warfare." Tim laughed (at least he knew what I was talking about).

"True, but he was so disconnected. His letter is not just coy, it's downright cryptic. How could any woman put up with that kind of other-worldliness for more than a week? My point is that he was on his own planet. No one else really mattered to him. He was alone. He loved being alone. According to him, and he said it countless times, the artist *had* to be alone."

He paused and smiled slyly. "And on Manitoulin Island, you know, he used to sit on rocks for hours and sing to cows."

Admittedly, this seemed persuasive. There is even a famous photograph of Gould sitting on Manitoulin Island in his trademark flat cloth hat, his knees drawn up as he lounges on a large rock and sings to a group of bemused cows. At least the captions say that he is singing to them; there would be no way of proving that, would there?

Downstairs, we went to visit Gould's piano, the immortal Steinway 317 194. It was built in New York in 1945 and was later known as CD318, because it's a model D. CD318 is part of the fabric of the Gould legend.

Pianists come from all over the world to see it, touch it, play it. It was for Gould's exclusive use from 1960 onward, and has a keyboard uncannily similar in feel and spacing to the Chickering he played as a child—which was why, he said, he loved it. For Gould, the keys of a piano had to be spaced exactly like those of a Chickering—it was one reason that he so detested harpsichords.

Today, CD318 is housed in a hall of gold mosaic columns, somewhat reminiscent, in its vulgarity, of a prominent Beirut disco of the early 1970s. But what did this matter? I climbed up, opened the scuffed lid, and played a few chords. The action was light, feathery; the sound crisply precise. Gould used to order the stagehands to laboriously insert wooden blocks under its legs to raise it still further above his seat. Combined with the childishly low chair, this brought the keyboard practically to the level of his nose. Alone in the hall, we listened to the ghostly chords of Gould's ghostly piano, and I felt a little unnerved. For me, there is always something a little spooky about Glenn Gould.

"You know," Tim said at length, "Gould was an interpreter of genius, a virtuoso of genius. But we can't get away from the fact that his specialness is, in the end, all about his Asperger's. Gould is famous, because he was such an extreme eccentric, because he was so clearly not one of us." Whether we like it or not, a genius today, he went on—he meant a genius in the realm of something quite marginal, like classical music or the higher mathematics—can only be famous because of things that lie outside his work. Things that have to do with his personality, his behavior. "Would Einstein be a household name without all the apocryphal tales about his eccentricities? Would Gould?"

"Probably not," I agreed. In the culture of celebrity, the piano player must be larger than life. Until recently, probably the most famous popular pianists were Liberace, or Dudley Moore. Precisely for this reason, I had always suspected that Gould had perhaps hammed things up a bit. He had a sophisticated understanding of modern media, was a friend of Marshall McLuhan's, a passionate advocate of the coming electronic age; he must have known all along that his reputation for eating only arrowroot biscuits and wearing winter clothes all summer long were bound to make him conspicuous, even notorious, in an era where serious work interests only a small minority of people. Tim disagreed.

"Gould's whole personality was shaped by Asperger's. Without it, for example, I don't believe he would have been so obsessed with contrapuntal music. And Gould was perhaps the greatest interpreter of contrapuntal music that has ever lived. His ability to hear many voices simultaneously, with amazing clarity, is truly unique. Look at his repertoire: Bach, Schönberg, the Beethoven concertos. It's a contrapuntal repertoire, an Asperger repertoire!"

His argument is that Gould possessed an uncanny knack for instantaneously "seizing" the structure of complex musical pieces in their totality. "This was an autistic strength," he insisted, as he led me back upstairs to the library, where I had been scheduled to look through some boxes of Gould's personal papers and manuscripts. "The phenomenal ability to concentrate, to memorize, to conceptualize in certain ways . . . to me, it is autistic. But then, of course, so were his enormous problems. Asperger's *made* him a celebrated artist." Tim wove his fingers together and smiled, Chinese sage–style. "Well, I know that's a controversial claim."

Afterward, we went down to the main floor and wandered around Tim's latest project—a show devoted to the life and times of jazz pianist Oscar Peterson. Peterson's personality, Tim argued, was the polar opposite of Gould's, because Peterson engaged in arguments with the world at large, which were no different from anyone else's. Hence the exhibited letters from the African National Congress, proclamations about media racism in cigarette advertisements, and the like. Gould would have regarded this as vulgar hucksterism. Gould argued for the electronic media and an end to the concert hall, not for political causes. "It's a matter of opinion who understood the world around him more acutely."

The Music Division's library looks out over the river. An assistant brought me box after box of Gould manuscripts and photographs, as well as reams of microfiched pages from his private journals. One strange photo showed Gould fingering a piano in a Bahamas nightclub called the Malamute in 1956, with the caption "Malamute pianist." But the journals were of most interest to me, and I pored through them for hours. Gould's handwriting was an incredible scrawl, somewhat resembling Egyptian demotic and, for all intents and purposes, unreadable. Through this scriptural haze, I could vaguely make out lists of drugs, his physical reactions to same, his phobias. On page 72, dated September 23 through January 30, 1979, I detached the relatively clear words, "this was not vastly different from other neck-stability systems. . . ."

Neck-stability systems? For a moment I had a vision of Temple Grandin in her squeeze machine. Then came the starker phrases: "neck-head-body," "I'm a chronic late-comer," "the tempo should relax a little," and other fragments. There were endless lists of accounts figures, as if Gould had obsessed continually about his very comfortable finances (he was a stock market whiz). Here and there, the words "Steinway Report" peeped out. In the margins hung construction-like doodles, while sentence after sentence had been maniacally scratched out, with little boxed numerals etched in above—some of them containing phone numbers. I noticed "Wilford 397-6908" written out several times. Then there were entirely empty pages headed by the word "scene," as in Scene V, version II, followed by a page with a single phrase, something like "Housekeeping $30" or "Mail 10 items."

Sometimes the scrawling was so wild that it seemed deranged. Gould's obsessive hypochondria reinforced this impression of chronic derailment, revealing a slipping and sliding mind obsessed with medications, almost none of which were legible. Taken superficially and at first glance, the journals impressed me as a transcription of profound instability; but, of course, that was taking things at face value.

The next morning I crossed the river to the Museum of Civilization, where there exists a bizarre archive of Gould's personal effects. As an officially designated Famous Canadian, Gould has, I suppose, earned the right to a niche in the national treasures repository, but one has to ask why this archival commemoration of a pianist consists of things that have nothing to do with music, but with clothes. I was met downstairs by the director of archives, Carmel Bejean. Carmel seemed a little wary at first. As we were walking up to the warehouse, she explained that the Gould archive was frequently besieged by hordes of sensation-seeking tourists, especially Japanese girls who loved to fondle the Maestro's old woolen jackets.

"Maybe," Carmel suggested, "it's something sexual. It quite distresses me, to be honest. It doesn't seem quite seemly. So I'm sorry that I'm a little suspicious."

In the vaults, we set up a long table and then went among the shelves of boxes. Gould's clothes are ceremoniously wrapped in paper, and Carmel laid out one item after another: the gray Keithmoor pants of virgin wool, the dowdy check Daks jackets, Harvey-Woods undershirts now sadly yellowed, a Schaffer-Hillman coat of Crombie's scotch fabric with a button missing and dog hairs still adhering, a hand-woven Donegal tartan hat with a guarantee and salute from one Paddy Ward sewn into it—*Joy and Health to you who wear this*. There was something squirey and old fogyish about Gould's wardrobe, if this was indeed a representative cross section of it. The Peter Scott wool cardigans could have handsomely draped my grandfather, as could the navy cotton shirts and pants, all bought at the Toronto store called Ely. The yellow Viyella shirts were another matter, but, all in all, Gould seemed to have dressed like a small-town bank manager with the occasional expensive taste. We moved back to the shelves that held an array of objects which had been removed from his house after his death. I picked up a heavy-tooled notebook bearing the words "The Essence of an Enigma" and opened it up. The pages were blank. Next to it stood a small collection of china dogs, mostly collies, and a note from his nephew, who had sent him one for Christmas 1974. "Dear Uncle Glenn, I hope you like it."

"He collected them," Carmel said.

We were both practically whispering in the huge room, which looks and feels like a top-secret aircraft hangar.

"What else did he collect?"

"Everything you see here. Wallets, for example."

There was indeed a veritable haul of leather wallets.

"And cuff links. He was crazy for cuff links."

The cuff links certainly came in dozens of shapes and sizes. I rummaged through them but found nothing remarkable. Carmel picked up an alarm clock and tilted it back and forth. "And these. He must have been a bit of a maniac about time."

Looking at the Westclox alarms, the Timex watches, and Taylor barometers, the compass set with ruler and triangle, I began to feel how pathetic are the mass-produced objects once owned by an individual now dead. Above the timepieces, I found an ancient knapsack that Gould had lugged halfway around the world as a touring concert pianist, with a ring of old Delta airline ticket stubs stamped at New York's Idlewild Airport. He hoarded these ticket stubs just as he hoarded china collies and hotel keys.

"Ah, the hotel keys!" Carmel shook her head. "There probably wasn't a single hotel anywhere in the world from which he didn't keep his room key. He was totally fascinated by those keys." Indeed, there was an impressive pile of them. "But did you see the wax finger bath?"

This was a portable machine that Gould used to give his hands a hot wax bath. We lifted up the lid and saw that there was still some solidified wax inside it.

"More famously," Carmel pointed out, "Gould was known for soaking his hands in hot water before a performance. The wax is less well-known."

I asked her if any other Famous Canadians had collections of posthumous clothes and bedside toys deposited in the Museum of Civilization. She had to admit that the Gould trove was somewhat unusual. We walked through darkened rows of antique Huron masks and feather capes, nineteenth-century carpets, and Japanese dolls. Perhaps, she offered, it was because Gould was famous for being odd, rather than for anything he actually did; of course, people loved his recordings. It wasn't that the recordings were beside the point; they were very much the point. But people found Gould haunting because of his personal peculiarities, almost as if he had been an idiot savant—in his case, of course, not an idiot savant but an Asperger's one. No doubt, she added, if they had a tin of those arrowroot biscuits, they would have filed them alongside the Viyella shirts. Gould's clothes and belongings were almost like those entombed with a pharaoh: they proved something crucial about him and told us something about his daily life.

"I mean, as Canadians we can't imagine Gould without his cloth cap, without the winter clothes in the heat of summer. Gould for us is a kind of performance artist who used real life as his stage. He is fantastical. That's the way he has to be. If he is not fantastical, he is not Glenn Gould."

"Did you ever think that he might have faked it?"

"Some people have suggested that. But no, when you look at the things we have here, you realize that he was genuinely strange."

"If he had not been strange would his cloth cap be in the Museum of Civilization today?"

Carmel laughed and then quickly looked over her shoulder. The warehouse is a little unnerving, as if the dead are listening to comments about them after all.

"I don't know," she said. "It's possible. It sounds crazy to say it, but it's possible. Who knows why people are famous? It's an irrational thing."

It is, I agreed, an irrational thing.

"But I like to think," she quickly added, "that he would have been famous anyway. What about those *Goldberg Variations*?"

★

Glenn Gould was born in Toronto in 1932. Originally, the family name was Gold, not Gould, and his father ran a successful furrier business called Gold Standard Furs. A possibly Jewish ancestry has been mooted, but Gold Senior, Bert, described himself simply as "English and Scottish." When asked, Gould himself exclaimed, "What? Me? With the name Gould and a father who's a furrier, and you're asking me if I'm Jewish?" His grandfather, one Thomas Gold, became alarmed at the number of furriers named Goldstein, Goldman, or (why not?) Goldberg, and decided to make Gold into Gould. He himself was the son of a Methodist minister, while the Scottish part of the family claimed the Norwegian composer Edvard Grieg as a relative. Gould was proud of this Scandinavian musical connection, as was his stern Presbyterian mother, Florence. Moreover, she was musically gifted as well as headstrong and deeply religious. Her musical energies went into church celebrations, and young Gould accordingly began his public keyboard life at the controls of a church organ.

Do the childhoods of geniuses tell us anything about them? The expatriate German psychiatrist Peter Ostwald, who befriended Gould later in his life via a passionate amateur interest in music, tried in his 1997 biography *Glenn Gould: The Ecstasy and Tragedy of Genius* to piece together Gould's childhood in the best psychoanalytic manner. Interviewing Gould's father, he asked whether little Gould had been "lusty." "He was reasonably lusty," Bert replied laconically. "But something unusual about him struck us from the beginning. When you'd expect a child to cry, Glenn would always hum. I think it was something in his makeup that made him hum rather than cry." Ostwald moves in for the interpretive kill. "Humming," he writes, "is a pleasure signal, soft and musical, whereas crying, which is louder and more noisy, indicates distress." Was it a sign of inborn musicality? Gould became noted for his constant, sometimes irritating, singing at the piano, and his father remembered that he did this from day one. Gould's tone-oriented mind seemed to have emerged intact almost from the womb.

But when it comes to genius, the laws of hagiography are difficult to suppress altogether, even with a chronicler as agile and aware as Ostwald. It is too tempting not to see significant meanings in those baby hummings. Bert recalls for Ostwald Gould's little fingers, always moving as if over an imaginary keyboard, even only hours after his birth!

> When Glenn was three days old his fingers never stopped moving, just like this, as if he's playing a scale [the father demonstrates by wiggling his fingers]. His arms would be swinging back and forth, and his fingers going. It showed us that Glenn was musical. And the doctor said, "That boy is going to be either a physician or a pianist— one or the other."

It was clearly a favorite family story. Would it have been told, however, if Gould had grown up to be a famous football player, or an unhappy accountant?

We can see that it's part and parcel of the myth of genius that the latter is seen as a creature perfectly crystallized almost before birth. In this respect, the genius and the autistic have much in common. And as it happens, Ostwald simultaneously turns his attention to precisely this dimension of the baby Gould.

Absence of crying, he decides, "is distinctly abnormal." Gould's mother was in her forties when she had him, apparently opening up yet another possibility of abnormality in the child. Then there was infant Gould's hand-flapping, associated with "peculiarities in speech development." This, to Ostwald's trained medical eye, is "suggestive of a developmental disorder called infantile autism." But . . . "Glenn obviously did not suffer from this disease."

> Had he been autistic, the remarkable success he had in a public career would have been impossible. But some of the behavior he manifested later in childhood and during his adolescence—a marked fear of certain physical objects, disturbances in empathy, social withdrawal, self-isolation, and obsessive attention to ritualized behavior—does resemble a condition called Asperger Disease, which is a variant of autism.

Ostwald then notes that Asperger Disease (a phrase I had not heard before and which I doubt is very popular today) is sometimes associated with unusual gifts in mathematics, music, drama, athletics, or art. He mentions Béla Bartók and Ludwig Wittgenstein as possible Asperger geniuses, but no athletic stars. Indeed, I cannot think of any great athletes who have been reputed to suffer from Asperger Disease, and since Ostwald is now unfortunately deceased I cannot call him up and ask him who he had in mind.

The musical psychologist then makes an interesting point. Gould, he points out, did indeed fulfill his family doctor's prophecy. For not only did he become a pianist, he also became a kind of expert, self-medicating hypochondriac who, apart from reading voraciously in medical literature, was also continually diagnosing and medicating himself. There are some hilarious letters from his youth in which Gould letterheads himself as "Gould's Clinic for Psycho-Pseumatic [sic] Therapy." In one of these he wrote to a fellow pianist:

> I am delighted to hear that Dr. Gould's perscriptions [sic] as usual proved efficacious. Due to my long experience with internal medicine practise I am unusually alert to the problems of neurotic artists. Whenever you are planning a trip up to Canada my nurse will be glad to arrange an appointment.

He then gaily suggests a regime of Nebutol and Luminal, sedatives that he himself used regularly.

Were Gould's eccentricities meaningful? Although Ostwald raises the question of Asperger's right at the beginning of his biography, he never returns to the subject subsequently. We are left merely with the unfolding spectacle of these oddities. The most iconographically potent of all of Gould's oddnesses were undoubtedly his overcoat, mittens, and cloth cap, all of which I had seen in the vault of the Museum of Civilization. As far as I can tell, the cloth cap first appears in the photographs after Gould's retirement from the stage in the early 1960s. In the '50s, Gould is extraordinarily dashing; Leonard Bernstein, who worked with him, tells an anecdote of how the usually disheveled Gould was once coifed by Mrs. Bernstein and emerged from their bathroom looking like a voluptuous saint. Bernstein was stunned by his unearthly beauty. And in the images of that decade there is no cloth cap in sight. The young pianist is elegantly brooding. Then suddenly the overcoat and cap appear as Gould poses at his cottage on Lake Simcoe. The

'60s?—one thinks. In many ways Gould was a very '60s figure, a kind of Sergeant Pepper imp. But he was stuck in the stuffy world of classical recitals, a world that he himself thought was dying, giving way to the new electronic age. As Gould retired from it, his dress changed. He ceased looking like a concert pianist and began gradually to assume the appearance of a dirty old man in a magazine store. Was this deliberate, a stage-managed rebellion against his professional milieu and its outdated hothouse personae—the Grand Artist in the style of Liszt—or was it the involuntary manifestation of his Aspergerish "disease"?

When I pore over these pictures of Gould posing in railroad cars or in summer fields in his thick woolen garments, I think back to a woman I once met in Singapore who dressed in exactly the same way and was said to suffer from an obscure mental illness called "frigophobia." The Chinese themselves call this bizarre mental disorder *wei han zheng*, or "fear of being cold." Sufferers become obsessed with heavy woolen clothing and begin swaddling themselves in thick cardigans, sweaters, woolly hats, and gloves. A Chinese specialist named Dr. Ng Beng Yeong explained to me that frigophobia was perhaps a psychotic response to the sudden rise of socially acceptable air-conditioning. "But really I haven't a clue," he admitted gaily. "Frigophobia doesn't even exist in the literature yet."

★

Gould's childhood neighborhood is called The Beach and borders Lake Ontario. It's a place of rolling hills and twisting semirural roads, the perfect Anglo-Saxon suburb: pious, unheroic, in Good Taste—whispering unheroic pieties. Gould's mother was herself a perfect avatar of this environment. "Florence Gould," wrote Gould's childhood friend Robert Fulford, "was a woman of propriety; when she spoke it was from a tranquil world of rules and order, a world from which conflict and tension had somehow been erased. She hated conflict, and she hated anything extreme or eccentric. . . . She longed to see her son have a 'normal' childhood. . . . In retrospect it's occurred to me that my friendship with Glenn perhaps owed something to Mrs. Gould's view that I was appropriately normal." It was a Christian home, he went on, "in which swearing of any kind was a grave effrontery. . . . Alone among all

my male contemporaries, [Glenn] never told dirty jokes, never specu-
lated about the sexuality of girls, and never said 'fuck'."

One can easily imagine the kind of respectable boys and girls who
lived in these modest mansions in the 1940s, entering fierce classical
piano competitions with the same regularity with which kids now go to
ecstasy raves. There were many child prodigy pianists in this milieu,
some of whom provided Gould with early competition. Self-con-
sciously, he outstripped them all. I remember the same thing from my
own rural town of Haywards Heath. It was a competitive social hysteria
based on classical music, with favored sons and daughters of the struggling but obscure upper middle class vying for supremacy in many a
drafty school and church performing hall.

With the decline of classical music's prestige, a decline that Gould
was all too aware of, this phenomenon has perhaps receded as well.
Winning piano competitions does not seem to matter much to adoles-
cents now. Nor are pianists sex symbols to them, in the way that the
smolderingly doomed Dinu Lipatti was to our parents or grandparents.

Gould seems to have sniffed the cultural decline of classical music
in the air, and he reacted vigorously to it. One way to do this was to act
the buffoon, and one should not forget that such a rebellion made big
waves in the relatively small and claustrophobic world of professional
music. Gould never dressed up like Liberace, but his dramatic refusal to
give concerts at all provoked the same scorn among the serious artists.
Ostwald sees it as a return to infancy, a retreat "to those peaceful, isolated organ lofts where he had practiced as a child." Gould himself
wrote:

> I discovered that, in the privacy, the solitude and (if all Freudians will
> stand clear) the womb-like security of the studio, it was possible to
> make music in a direct, more personal manner than any concert hall
> would ever permit. . . .

Music as a public spectacle made no sense whatsoever to Gould. It
was a contradiction in terms. He loved recording technology precisely
because it was estranging—for how could music exist without loneliness? "The greatest of all teachers," he once remarked, "is the tape
recorder. I would be lost without it."

This love of isolation hung around Gould like a nimbus: he was always alone. Interestingly enough, when he was a child Gould also loved to disappear for endless voyages on his bicycle. "He'd strike off on the bicycle," his father recalled, "and his mother would get a little anxious, wondering where he was . . . and I'd take the car and maybe find him five miles away on the side of the road. And one day I came along and found him singing to a bunch of cows. They were all lined up inside the fence."

Other times he would escape in the family motorboat, roaming as far as fifteen miles away and returning home with his hands conducting inaudible symphonies. In common with many Asperger children, he was critically anxious and often expressed his anxieties in curiously mathematical forms.

The journalist Pierre Berton, in an interview with Gould in 1959, described the young Gould's terror of humiliation after the latter had seen another boy fall ill at school in front of all his classmates:

> All eyes turned on the wretched child and from that instant on Gould was haunted by the specter of himself being ill in public. That afternoon he returned to school with two soda mints in his pocket, a small tousled boy on guard against the moment when he might lose face. The soda mints were soon supplemented by aspirins and then by more pills. In school, Gould literally counted each second until lunch hour (10,800 seconds at 9:00 A.M., a comforting four-figure 9,900 at 9:15), and prayed that nothing might happen to humiliate him.

The breaking of time into seconds, the smallest reasonable units for the clock-conscious, is quite common among Asperger boys, for whom this deconstruction of hours, days, or even weeks into seconds is a soothing comfort. So is a love of animals. One of Gould's childhood dreams, according to Ostwald, was a plan to create a paradisiacal old-age home for stray animals situated in the great Canadian North, specifically in Manitoulin Island, a place said to be the dwelling of the god Manitou. Gould kept four goldfish named Bach, Beethoven, Haydn, and Chopin. Once, after capturing a skunk, the young boy wrote, "I am a skunk, a skunk am I. Skunking is all I know, I want no more. I am a skunk, a skunk I'll remain. . . ." (I think back to the sad dog hairs plastered all over his cataloged jackets and coat.)

Then there were aversions. He hated bright colors. "I wouldn't have, as a child," he later wrote, "any toy that was colored red at all." And he added, as if to illustrate where this aversion to redness had led: "I hate clear days; I hate the sunlight; I hate yellow. . . . To long for a gray day was, for me, the ultimate one could achieve in the world." Ostwald suggests that being black and white—monochrome in other words—was one of the piano's most powerful attractions for Gould. It was an island of anti-color.

In 1946, when he was fourteen, Gould experienced an offbeat turning point in his musical evolution. One day, while he was playing a Mozart fugue, the maid turned on a vacuum cleaner close to the piano. The mechanical buzz of the machine suddenly drowned out the sound of the piano, but in a way that the player found intriguing, even pleasant. Gould later compared the effect to singing in the bath with both ears full of water while rocking one's head from side to side.

But as the sound of the piano diminished, Gould's tactile sense of his own pianistic movements intensified. "I could imagine what I was doing, but I couldn't actually hear it."

"What had happened," Ostwald writes, "was that the masking noise of the vacuum cleaner had shifted Glenn's attention to the internal sensations of his body and away from the acoustical results of his playing. It was like a trip to the interior—and he enjoyed it."

Ostwald compares this epiphany to that undergone by other artists who experienced new dimensions of themselves due to similarly accidental events—Beethoven, for example, overwhelmed by his deafness. Inner hearing, in other words, overwhelmed outer experience.

"The strange thing," Gould mused, "was that all of it suddenly sounded better than it had without the vacuum cleaner, and those parts which I couldn't actually hear sounded best of all."

Thereafter, when learning new scores, he would turn on a TV set, a Beatles record, or a radio in order to drown out the sound of what he was playing and send him off into the realm of the "inner ear."

In later life, this disconnection between inner and outer gave Gould an unearthly quality, which his biographer captures neatly in a small vignette. After an evening of impromptu music-making, Ostwald drives Gould back to his hotel:

As we drove there, I noticed a peculiar quality of detachment and isolation. Despite his overt friendliness and jovial humor, Glenn radiated little warmth, almost as if the bodily coldness he often complained of had chilled him spiritually. He spoke of music but said absolutely nothing about the musicians we had just been with, neither their personalities nor their performance, nor did he have any comments whatsoever, positive or negative, about my violin playing. It suddenly occurred to me that during the five hours we'd spent together, Glenn had minimized all human relations; he'd said very little about his family and almost nothing about any friends, teachers, or other people who might have been close to him. The talk had focused primarily on himself, his musical activities, and his love of animals.

On one occasion, Gould casually remarked, "People are about as important to me as food." This from a man who mostly ate scrambled eggs and arrowroot biscuits, usually at 5 A.M. in all-night diners. He called his rudimentary meals "scrambleds."

In the late 1960s, Gould began voyaging into the Canadian North, taking the train as far as the remote town of Churchill or staying in motels around Lake Superior in places like Wawa and Marathon. Isolation and ecstasy: one produced the other. Here he devised a revolutionary radio documentary called *The Idea of North*, a series of interviews with four unrelated people about their experience of the North.

Gould quickly experimented with simultaneous dialogues, juxtaposing and overlapping the four voices so that sometimes they talked at the same time instead of in a linear sequence. It worked in much the same way that musical voices are sewn together in a fugue, a contrapuntal method that suited Gould well. Indeed it was at the core of his sensibility, both in his music and in his pathology, for he would sometimes hear imaginary voices whispering alongside real ones during ordinary conversations. He called it "contrapuntal radio" and thought of it as possible new art form.

These strange documentaries were run on CBC radio and often dealt with solitary people (other well-known episodes included "The Latecomers" and "The Quiet in the Land"). They are difficult to describe. *New York Times* writer Robert Hurwitz claimed that they were "comparable to sitting on the IRT during rush hour, reading a newspaper, while picking up snatches of two or three conversations as a portable radio blasts in the background and the car rattles down the track." "The Idea of North," for example, begins with a woman's voice muttering about seals, polar bears, frozen lakes, and snow. Then two other voices join in, creating a virtually incomprehensible vocal tapestry. After they fade away, Gould's voice comes in to make his introduction, quickly followed by the sound of a departing train. Gould himself regarded these sound-tapestries as music. "It really is, in fact, composition," he said in a 1970 TV interview.

Quite apart from its artistic possibilities, though, contrapuntal radio also corresponded with Gould's need to invent characters for himself, "for externalizing aspects of his contrapuntal mind," as Ostwald has it. Gould often referred to himself in the third person, in the way that boxers sometimes do.

". . . my most joyous moments in radio," he confessed, "as opposed to my most creative ones, perhaps, are those when I turn to impersonation. I was incapable of writing in a sustained humorous style until I developed an ability to portray myself pseudonymously."

Gould had a stock of Dickensian alter egos that he used to "pontificate," as he put it, on North American culture. "One for every season," as he used to explain. One of them was Herbert von Hochmeister, a parody of Herbert von Karajan, a retired conductor who liked to spout flamboyant cultural theories. Others included Sir Humphrey Price-Davies (who liked to cock his nose at the "extravagantly eccentric Canadian pianist Glenn Gould"). Then there was a quack psychoanalyst named S. F. Lemming, M.D., a Marxist critic called Zoltán Mostyani (*Journal of the All-Union Musical Workers of Budapest*), and the doddering Sir Nigel Twitt-Thornwaite, for which Gould dressed up in stand-fall collar and white moustaches—"the epitome of Edwardian mod." There was even Myron Chianti, a version of the young Marlon Brando, who stumbles around muttering louche inanities (but Gould admired Brando, a type diametrically opposed to himself).

Impersonations, costumes, masks: they are the lifeblood of certain artists whose personalities divide voluptuously into private and public spheres. Gould was like Oscar Wilde in that personas allowed him to engage in cultural warfare. While silk suits and dyed carnations allowed Wilde to create a public homosexual intellectual, dowdy cloth caps and mittens allowed Gould to create a crabby anti-artist with a taste for denunciation.

But Gould's witty dialogues with himself, in which he interrogates himself, destroys his own taboos, and mocks himself, are not nearly as clinically narcissistic as some commentators have suggested. They suggest only that he did have inner voices. Many of them go something like this:

> G.G.: May I speak now?
> g.g.: Of course, I didn't mean to get carried away, but I do feel strongly about the—
> G.G.: —about the artist as superman?
> g.g.: That's not quite fair, Mr. Gould.
> G.G.: Or as the interlocutor as controller of conversations, perhaps?
> g.g.: There's certainly no need to be rude. I didn't really expect a conciliatory response from you—I realize that you've staked out certain philosophical claims in regard to these issues—but I did at least hope that just once you'd confess to a personal experience of the one-to-one, artist-to-listener relationship, I had hoped that you might confess to having personally been witness to the magnetic attraction of a great artist visible at work before his public.
> G.G.: Oh, I have had that experience.
> g.g.: Really?

Gould's philosophy? Geoffrey Payzant, author of another biography, claims that Gould had tried to separate music from cruelty. Gould loathed the very principle of competition, even the competition between soloist and orchestra in a concerto. For him, technology was what broke up the natural laws of survival and took the red out of human tooth and claw. Of course, like most people who detest competition, Gould was ferociously competitive himself. He moralized against capitalism, while winning huge amounts of money on the stock exchange. But his deeper refusal of competition was naturally expressed in his solitude, his championing of solitude. The question for the diagnosers of Gould is whether this extreme and earnest quest for solitude is "normal," or whether it reveals a mental illness—a nuanced form, but mental illness nonetheless.

★

Several doctors have come forward to denounce the Asperger's diagnosis of Gould. Among them are psychologists Helen Metaros and Lynne Walter. Metaros, a Professor at the University of Toronto, offers her own diagnosis: "His current and rather complex diagnosis would be recurrent depression overlapping with social phobia and with obsessive and unrelenting hypochondria."

Walter, for her part, is troubled by the prospect of a brilliant man who seemed to be, as she put it, "notably eccentric." In a paper published in *Glenn Gould* magazine, she notes that Gould referred to himself as the "Last Puritan." Walter is puzzled. What could he have possibly meant by such a phrase? Moreover, he seemed to have been actually proud of this designation. Walter even suspects that Gould's extreme moral posturing was a part of his illness. "Indeed," she writes, Gould's "obsessional, schizoid and narcissistic qualities all contributed to his moralistic attitude: his Puritanism crossed the spectrum of those conditions." What she seems to mean here is that Gould's austere and solitary life was the result of a mental disorder of some kind, and his so-called Puritanism (by which Gould meant his withdrawal from ordinary pleasures and from ordinary life) was therefore pathological.

It seems impossible to a contemporary psychiatrist that Puritanism, even the obviously tongue-in-cheek variety espoused by Gould, could be a moral choice that a sane individual might just happen to elect. No, it has to be a sign of illness, a disorder, a psychiatric "condition." Walter sums Gould up in no uncertain terms:

> Comprising Gould's narcissistic traits were his impenetrability and tendency to self-reference; his rejection of interpretations or criticisms by others; his low capacity for empathy; his pride in independence; his grandiosity; his expectation of being entitled to others' services; and his haughty attitude. Some regarded him as exploiting people for his own purposes, and dropping them when they were no longer useful.

It's interesting to see how psychiatry regards things such as "pride in independence" or "grandiosity" as faults, whereas in other circumstances (and in other ages) they would be regarded as virtues or the concomitants of virtue—though not, of course, by the Puritans! Is it because this same discipline cannot really digest or accept exceptional creative personalities—indeed, implicitly loathes them? Walter's conclusion is sonorous. "The most appropriate diagnostic category," she writes firmly, "seems to be Mixed Personality Disorder."

I have to admit that when I read this I fell off my chair. Mixed Personality Disorder? Elsewhere, she suggests that creativity is itself "schizoid," which may be true. But then again, what model of normality could be worth anything that does not include it? As for Mixed Personality, why would this be considered undesirable, let alone a disorder? What, in any case, is an Unmixed Personality?

Gould's rejection of external interpretation and criticism of himself seems to me essentially sane. He disliked psychoanalysts. He accepted, for one thing, that he did not know himself; how therefore could he be known by a stranger? Maybe it's this emphatic insistence that human personalities cannot ultimately be known that makes psychiatry uneasy, even annoyed. It's easy to admit unknowability intellectually, but far harder to act accordingly.

Other followers of Gould have posted their own retaliations to the presumptions of psychiatry on various websites, especially one called *glenngould.org*. I found there some virulently spirited comments from a Gould fan interestingly named Elmer Elevator. Mister Elevator tore into the very existence of Asperger's itself:

> Again I would remind people of the notorious trendiness of psychology and psychiatry. Will Asperger's make the time cut? Will anyone even remember it twenty years from now, except as an embarrassing footnote?

"Who," he thundered, "has bedroom wall posters of Asperger?" And he continued:

> I think Asperger's Syndrome, certainly as applied to Glenn Gould, is a symptom of our increasing and widespread dread of non-conformity, and trying to stamp out non-conformity and diversity is a disease psychologists have always suffered from. Most of us have made our painful adjustments to the cookie cutter, and deeply resent and fear anyone who notoriously and unashamedly resisted being mass-processed. . . .

Gould played charades with his own neuroses, which were part of his musical drives, and toyed with his interpreters, perhaps even a little too obviously and deviously. But his contempt for some of them was intuitively vivid. Inversely, Gould has become something of a folk hero to the Asperger world, and one has to ask whether this doesn't have something to do with his gay and witty "up yours" to the establishment. To be alive, he tells us, is to disengage, to evade the cookie-cutter, as Elmer Elevator has it. Gould thus upholds the ancient tradition of the genius as lunatic, as social idiot, as idiot savant, and therefore as abnormal exception. But he may well have been not only the last so-called Puritan—he may have been the last genius as well.

CHAPTER 4

RAIN MEN

Genius may be an abnormality.
—Temple Grandin

It was 9 o'clock at night by the time I got to LaGuardia to meet Jerry Newport. It was winter, the airport looked tired and half-deserted, the dense gloom of the low corridors and escalators touched by a scent of snow. By Terminal C, the retail outlets were even grimmer than they are by day, and exhausted Mexican staff were closing up the Sunglass Hut with elaborate rings of keys that made them look like guards in some desultory tropical jail. I went to the Wendy's and sat down under its undulating strawberry neons. Jerry Newport had not told me where exactly to meet him, merely that his fiancée would be arriving at Gate 4 from Houston and that I was to meet them both there. But what did Jerry Newport look like? And what did the fiancée of Jerry Newport look like? Jerry Newport was as much a mystery as his fiancée, and I may add that it was a further mystery to me why Jerry Newport had insisted that we meet inside LaGuardia airport at 9 o'clock at night. But he *had* insisted, saying with admirable certitude, "I'm afraid I can only be interviewed in an airport. Sorry."

I ordered coffee after coffee, looking constantly at my watch and wondering why Jerry Newport was late. For after all, as I said, it had been his idea. The arrivals terminal that he had suggested for a rendezvous looked like a pick-up zone for Cold War agents. Eventually, since I had no idea what Jerry Newport looked like, I wandered "with intent" (as the police have it) around the Sunglass Hut and the airport

Wendy's, looking for suspects. To tell the truth, I was a little resentful because although, as I have said, I adore airports, I cannot adore LaGuardia. It's a cramped, claustrophobic warren—exactly the kind of airport in which you can imagine terrorists settling in for some serious fun. No one looks quite wholesome inside LaGuardia. Faces become sallow, moods sour into anxiety. There is an underdevelopment to the texture of things, a nightmarish light like Greek Fire. There is that shifty energy that one finds in large bus stations like the Port Authority late at night when the laws of civility are wearing thin. Lonely figures, not quite bums, men waiting alone with little paper pots of ketchup and baskets of onion rings while Africans sweep the floors.

Under the phosphorous glare a single man sat hunched over a red tray, but it was not Jerry Newport. An hour went by and I felt like tapping this solitary diner on the shoulder anyway. I felt like asking him what the hell he thought he was doing eating off a red tray at the LaGuardia Wendy's. Instead, I sat down as far away from him as I could and waited for Jerry Newport. Another half-hour went by. I began to scrutinize passing faces to see if any of them could be called Aspergerish. But perhaps I should explain who Jerry Newport is.

Jerry Newport is the most famous Asperger savant in America. Although I had never met him, I had seen him once on an edition of *60 Minutes*. He was a chirpy man, on the edge of middle age, who made mathematical calculations at the speed of light. Perhaps it was the fault of a highly unreliable memory, but I seem to recall Jerry gaily dueling with a Princeton math professor and winning handily.

Nevertheless, I had no real idea who he was. Was he like the original Rain Man? Could he, too, calculate the precise number of a fistful of toothpicks dropped haphazardly on the floor? For a while I heard his name from afar. It would crop up on websites or in the middle of incongruous conversations on unrelated topics. He was a little like Darius McCollum, an Asperger fugitive whose claim to notoriety was not quite clear but who seemed to loom large somewhere in what could be called the Autistic Underground. Jerry was a genius; Jerry was a rebel. But did genius Jerry have a cause?

Jerry was a very modern kind of celebrity, a man celebrated more for his abnormality than for his actual gifts. He existed in a medical twilight zone. When I finally called him at his home in Los Angeles, I found myself hopelessly entangled with a piping, excitable voice, which

I imagined attached to a childlike body and a menacing grin. Entangled is the right word, because I felt at once that conversing with Jerry was like struggling with an optimistically tireless anaconda that never lets go. It was the Agony of Laocoön all over again. In the background I could hear a cackling and cooing of what sounded like an inordinate number of cockatoos. "My parrots," Jerry said. "It's what I collect."

Over the next two weeks, we played cat and mouse by phone. Jerry was working for the accounts department of a Los Angeles telephone company but always appeared to be on the move. He was speeding between airports, driving hither and thither, rushing from point A to point B. Sometimes he would phone at five-minute intervals as he took cabs here and there. "Hi, this is Jerry. It's 5.34 and I'm heading downtown on Interstate 495!" Then, "It's me again, I'm heading southeast on Interstate 5, it's 5.43!"

As I've already said, I have an affinity for airports, and it made me a little uneasy that Jerry appeared to like them as well. Could the roots of our respective affinities be related? At last, however, a beaming, rather roly-poly man came hurtling out of the gloom in a racetrack T-shirt that bore a picture of galloping horses and the words *Let It Ride!* He looked a bit like Jack Nicholson, with sandy hair and small tuft of hair on his nose. In one hand he carried a huge Willy the Whale soda cup with a molded orca rising from its lid.

His good-natured energy was obvious at once. It bounced around like a rubber ball, hitting everything and anything at will. But I noticed that, all around him, there spread a subtle, barely noticeable ripple of alarm. People gave him a swift second glance and then stepped a little to one side, as if he had winked at them out of turn—or as if that rubber ball had hit them square between the eyes. As soon as he saw me, however, Jerry knew who I was; he took a last sip on the straw protruding from the orca and wiped his hands on his T-shirt, which seemed already considerably used. He came up merrily, all handshakes and laughter.

"When were you born?" he asked at once.

I gave the date. "So," he thought for a split second, "that would be a Monday. Correct?"

He sat down and put Willy the Whale next to him.

"Correct," I said.

"Sometimes I get it wrong. But only by one day."

"Isn't Mary's flight in by now?"

"No, no, it's late." His large mass shook with another small tremulation of merriment. "Mary's always late."

This, of course, was a joke; he looked keenly to see if I got it. Then he added, "What if we multiply the year of your birth with the year of your birth?"

It took him three seconds.

"Naturally," I said, "I can't verify if that's the correct answer."

"You don't need to. It's correct."

"What if you multiply that figure by 16?"

Rolling his eyes, he reeled off a preposterously long number.

"That's an easy one. What if you'd asked me to do a fraction?" His eyes took on a brief naive malice. "Ah, that would have been a different story altogether. It would have taken me a whole ten seconds longer."

Jerry can apparently do these enormous multiplications in his head faster than any Princeton professor using a pencil. But he is also happy to explain exactly how he does it, though it takes him far longer to utter the explanation than it does to do the calculation itself. As he was explaining to me how exactly he had multiplied the year of my birth by the year of my birth and then multiplied that figure by sixteen, his eye wandered up to the ceiling fan above us. Its motion clearly attracted him. When I asked him if anything was the matter he stooped down to the soda straw and blinked.

"Nothing's the matter. It's just that I like spinning things. I used to be a cab driver, you know. And when I was a cab driver, I would go to all the truck stops and watch the wheels spinning. I love spinning hubcaps. Anything that spins gets me going. As a matter of fact, I only like bars that have fans. If a bar doesn't have a fan, I pass it by. There's nothing in the bar that's spinning."

Now I confess that I had not come to my interview with Jerry unprepared. There is an Asperger writer in Los Angeles named Jonathan Mitchell who writes short stories that are often on autistic themes. One of them is about Jerry and it's called "Guess Who Isn't Coming to Lunch." Jonathan had been kind enough to send me a copy of it, and I had read it carefully in the hope that it might be able to tell me something about Jerry, whom Jonathan admires and likes. The first paragraph was certainly appealing enough, with Jerry cast as an autistic character named Arthur:

Prime numbers and presidents. Petra had never dreamed that she would find these subjects sexy until she met and fell in love with Arthur. The way Arthur, her fiancé, obsessed over them was a turn on. He was mildly autistic and a mathematical savant. He was also obsessed with lurid facts about American presidents. She loved the way he could tell you what day of the week you were born on when you gave him your birthday, how he could do lightning-fast mathematical calculations, and how he rattled off the sleazier side of trivia involving every president of the United States from George Washington to Bill Clinton. Arthur also liked to look at license plate numbers and say (often out loud) whether the license plate was prime or composite. It was a nice pleasant mid April day. Arthur had also told her how many hours, how many minutes and how many seconds it had been since he had proposed but she had forgotten. What mattered was that she was in love.

Arthur had been the most exciting man she had ever met. Arthur had until March 28 to propose to her. After all, 28 was his favorite number. It was the perfect number—the sum of its factors equaled 28.

I had wondered to myself if a real Asperger's person would propose to a woman on a given day simply because the sum of that number's factors equaled the number itself. Was this a window into an emotional universe I could even imagine? But I couldn't quite summon up the courage to ask Jerry if he himself had done this with Mary. For one thing, I knew that they had been married and divorced once before and were now engaged for a second time, and given this circumstance it didn't seem polite or fair to ask. When Arthur brings her a dozen long-stemmed roses at the beginning of the story, he and Petra kiss twice. "Two kisses," Arthur says, "for each factor of the roses before you get to a prime number."

> "Did I ever tell you that the license plate number on your car 3NXP613, or 3,613, is a prime number?" asked Arthur.
> "Only about a hundred times," said Petra, smiling a bit.

A little later, as they are anxiously discussing how Petra will inform her parents that she's going to marry an Asperger's man, Arthur suddenly nervously exclaims: "I'm one of the handful of autistic men who has managed to find a wife. I'm not the kind of guy one brings home to one's parents." And, thrown into an anxiety fit, he begins jumping up and down while shaking his hands in front of his face and exclaiming

"283 times 396 equals 112,068!"—something, the author adds, he often did when he was nervous.

I was fascinated. Was this what Jerry's conjugal life was like? Pillow talk as number crunching? Again, it didn't seem possible to ask him upfront, but Jonathan surely knew what he was talking about as a fellow Asperger's man. As for Jerry, he was more comfortable talking about himself. His life, he said, had been something of a blank slate between the ages of twenty-two and forty-five. After graduating, he worked in a variety of meaningless jobs, as most Asperger's men do. In his twenties, he went for an interview in the Transamerica Building in Los Angeles. The interview, he related, went something like this:

> Interviewer: So, Jerry, what do you want to do?
> Jerry: Go to Hollywood Park and eat ice cream.
> Interviewer: Next!

"It was worse when I was younger," he said. "I hate being touched by anyone. And I'd curse when I was walking in the street. I worked in a library, but I was fired for talking too loud. Then I went to see *Rain Man*. But I thought that can't be me, because he's in an institution. And I'm not Rain Man. For one thing, I've got Mary." He brightened at once. "Shall we go look for Mary?"

We went to look for Mary.

As we made our way to the arrivals gate, I noticed that Jerry has an oddly rolling gait—a kind of feverish shuffling about him, which certainly catches the eye of the average neurotypical. Normality and its opposite, I realized, are subtly and instantaneously perceived on a very physical level. It must, I thought, be an instinct of some sort. A circle seemed to open up around us, as if we were at the center of a force field of animal electricity that set off primitive feelings of attraction and repulsion.

"When I think of Mary," Jerry was saying, "I calculate backwards."

Jerry and Mary have now moved to Tucson, where they not only keep Jerry's collection of fourteen exotic birds (parakeets, lovebirds, cockatoos) but also spend their time betting at the local racetracks—dogs and horses. Jerry uses his mathematical skills to work the odds.

"We do pretty well, all in all. We could almost make a living at it."

I asked him what else he did with his time, apart from working as a substitute math teacher in the Tucson public school system. He replied that he was writing a self-help book for Asperger's people. "I like doing it, because it's slow." The main thing in his life, I surmised, was meeting Mary at an Asperger's Halloween party.

He was now getting fidgety and anxious, as the passengers from her flight trickled through the glass-walled exit corridor. I wondered if he was going to start jumping up and down, crying, "283 times 396 equals 112,068!" Then there she was: a redhead in a kind of leopard-spot suit. Jerry pitched toward her, but on contact the hug he gave was perfunctory. Mary seemed a little dazed after her flight and shook my hand quietly. I offered to drive them back into Manhattan to their hotel near Times Square.

As we wandered around the airport parking lot, I admitted that I couldn't quite remember where I'd put the car.

"Then you *must* be one of us," Jerry kept saying. "I *knew* it, Larry!"

No one ever calls me Larry. It's an irritating name, a name I hate. But I let it go. Jerry broke into a demonically innocent cackle. He had suddenly become incredibly voluble, releasing a pitter-patter stream of one-sided dialogue, while Mary nodded and nodded with an expression somewhere between forgiveness and fatigue. The drive into Manhattan was a verbal riot.

"Einstein," Jerry crowed, "definitely one of us. Wittgenstein, absolutely. Bill Gates. You didn't know about Bill Gates? Big secret! I wouldn't be surprised if old Bill had it. One look at Microsoft—"

"What about Glenn Gould?" I said.

"Glenn Gould? Never heard of that one. Have you heard of that one, Mary? But the group Herman's Hermits is a possibility. Maybe they all had it."

"Herman's Hermits?"

"Yeah, don't you know Herman's Hermits? Herman's Hermits wrote the ultimate autistic song."

And suddenly Jerry and Mary began singing together, rocking themselves slightly from side to side.

> I'm 'Enery the Eighth I am,
> 'Enery the Eighth I am, I am.

"It's the ultimate autistic song," Jerry cried triumphantly. "Do you know why, Larry?"

"I have no idea."

"Because every stanza is exactly the same as the first. Isn't it so, Mary?"

"I love Herman's Hermits," said Mary intensely.

★

The term "idiot savant" was coined by the English doctor J. Langdon Down, the same man who gave his name to another and much more debilitating childhood syndrome. In his 1887 Lettisonian Lectures to the Medical Society of London, Down admitted that he had "no liking for the term 'idiot.' It is so frequently a term of reproach." Nevertheless, it had to be used in order to describe the baffling paradox of extreme debility and superiority existing within a single individual.

One of Down's patients had apparently memorized much of Gibbon's monumental *Decline and Fall of the Roman Empire*. On his first reading of it, however, he had made a tiny error, and, thereafter, whenever reciting it, he would always correct the same tiny error. Down called this remarkable facility for memory "verbal adhesion." It was, effectively, memory without consciousness. He also noticed a capacity for "lightning calculation," which is self-explanatory, and a refined obsession with passing time. One boy, who could not read clocks, could nevertheless report the time accurately, as if he had an organic clock ticking constantly inside him. Down also noted that savants were almost always boys.

The most complete survey of savants in English is the American psychiatrist Darold Treffert's *Extraordinary People*, which was first published in 1973. In this remarkable book, Treffert gives us a complete history of the scientific unraveling of the idiot savant, from the work of Down to his own research into savantism. While Down was puzzling over brilliant "idiots" in London, Albert Binet (the inventor of the IQ test) and Jean Martin Charcot had begun investigating savant intelligence at the Salpêtrière Hospital in Paris. It was Binet who first suggested differences between auditory and visual intelligence. And it was Binet who first pointed out that lightning and calendrical calculations might be linked. He also hinted that such calculation abilities might be performed by the unconscious. In an 1894 treatise, Binet concluded:

> The unconscious which is within us, and which psychology has in
> recent years often succeeded in illuminating, is perhaps capable of
> foreseeing the solution to a problem or long arithmetic operations
> without carrying the details of the calculations.

How savants do their calculations has remained mysterious. In *Thinking in Pictures*, Temple Grandin argues that savants can actually visualize long chains of calculations in the same way that experienced Chinese mathematicians can shift the balances of abacuses in their minds without actually moving the beads of a real one with their hands. Grandin claims that she can guess the unfinished part of, say, a jigsaw puzzle by using what she calls her mental "video library." This is a vast store of remembered facts that can be consulted visually, as if each piece of information were a kind of photograph or short film. Chinese mathematicians work in the same way:

> When a mathematician becomes really skilled, he simply visualizes
> the abacus in his imagination and no longer needs a real one. The
> beads move on a visualized video abacus in his brain.

At a 1909 meeting of the Society for Psychiatry and Neurology in Vienna, a Dr. A. Witzmann showed a twenty-year-old asylum inmate who could spontaneously calculate the day of the week for any given date occurring between the birth of Christ and 1909. Witzmann pointed out, however, that for reasons unknown, this man's knowledge of the calendar stopped abruptly in the year 2000. Recalling that neurologists had previously noticed persons who could easily reproduce vast quantities of statistics about railway timetables, budgets, and accounts books, Witzmann decided to call such people "brain athletes."

Five years later, in 1914, the British doctor Alfred Tredgold, a Consulting Physician of the direly named National Association for the Feeble-Minded, published the classic paper on idiot savants in a chapter of his book *Mental Deficiencies (Amentia)*.

In this paper, which is still a standard reference work in the English-speaking world, Tredgold reiterates that savants are almost always male but that their IQs rarely fall into the true idiot category of below 25. Some of his case histories were startling. There was, for example, a hypothyroid cretin ironically named Gottfried Mind, who excelled at making spectacular pictures of cats. Mind, who was born in 1768 and died at the age of forty-six, was illiterate and had remarkably huge and rough hands. Yet his pictures of cats became so famous that he became known as the Cat's Raphael and figured a startled King George IV among his appreciative patrons.

Another patient illustrated remarkable powers of savant memory: a fifty-six-year-old man who could remember everything pertaining to his own life, even down to the most minute technical detail, and who, as a result, became a valuable record of everything that had transpired inside his institution! Tredgold also cites several examples of calendrical calculators as well as examples of musical savants and superior calcula-tors, one of whom could instantly calculate how many minutes a person had lived. One rare female calculator could divide the number 576,560 by 16 almost spontaneously but could not do the most basic arithmetic. But perhaps Tredgold's most fascinating case history is a savant artist named James Henry Pullen, otherwise known as the Genius of Earlswood Asylum, after the institution in which he was confined.

Deaf and dumb from birth, Pullen entered Earlswood when he was fifteen and lived there for sixty-six years. "It is Pullen who comes to mind," Treffert writes, "when Dostoyevsky describes the idiot as a pri-vate, unrelated person, an outsider."

"Because of his deafness and muteness," Treffert goes on, Pullen "was isolated, eccentric and alone." He could hardly form sentences, crying in a mixture of monosyllables and hand gestures. Until he was seven years old, the only word he uttered was "muvver." But early on, he nevertheless became obsessed with model ships. He quickly became skilled at both carving and making intricate drawings of them. As he grew older, these works, often executed in dark chalk, decorated the gloomy corridors of Earlswood. One of them was a representation of the universe as a ship, with ivory angels hovering around it and a figure of Satan at the stern. At the age of thirty-five, Pullen embarked on his masterpiece, a huge model ship called *The Great Eastern*. It contained almost six thousand rivets and a vast number of planks, along with thirteen lifeboats hoisted on davits and fully decorated state cabins. After seven years of labor, this stupen-dous artifact won first prize at the 1883 Fisheries Exhibition in England, and Pullen became an instant national celebrity.

But Pullen was a dark character. Often wild and sullen, he was notorious for his unpleasantly unpredictable behavior. According to another nineteenth-century savant researcher, Dr. Edward Sequin, it took Pullen six months to learn the difference between a dog's head and its tail when he was a child. He never learned to read or write, and as he aged he became more morbidly suspicious, morose, and even violent. On one occasion he constructed a guillotine-like contraption over his workshop door, in the hope of injuring a colleague whom he disliked. He also constructed an enormous mannequin inside which he could sit, control its arms and legs, and communicate through a bugle attached to its mouth. And because he was unusually sensitive to vibrations coming through the ground, he devised an alarm system in his workshop based on that sensitivity. It made him aware of any approaching visitor. Tredgold concludes:

> His powers of observation, comparison, attention, memory, will and pertinacity are extraordinary; and yet he is obviously too childish, and at the same time too emotional, unstable, and lacking in mental balance to make any headway, or even to hold his own, in the outside world. Without someone to stage-manage him, his remarkable gifts would never suffice to supply him with the necessities of life. . . .

It seemed to Tredgold that Pullen's isolation had caused all the faculties of his mind to pour into one narrow, focused channel. This focus was what made him brilliant at one thing, namely building model ships. Others, however, pointed out that his defects seemed to run deeper. The contemporary doctor F. Sano observed that Pullen, "with both his eyes wide open to the bright world of London, and his skilled ten fingers under complete sense control, even after having been busy for months in the printer's shop at Earlswood, could not absorb, digest, or exteriorise the most ordinary sentence of politeness. To say, 'I am very much obliged to you' was strange to him in grammatical arrangement as well as in social meaning."

It would seem that Pullen suffered from a social illiteracy or Mindblindness, which brings to mind that of nondeaf autistics. His illness was more profound than an isolation brought about by deafness. Sano went out to perform a postmortem autopsy of Pullen's brain, one of the most detailed dissections of a savant brain ever accomplished, the results of which were published in the *Journal of Mental Science* in 1918. A slightly overlarge corpus callosum—a bundle of nerves connecting the right and left hemispheres of the brain—and traces of arteriosclerosis seemed to explain at least some of Pullen's abnormalities as well some of his abilities, but nothing was proved conclusively.

Savants, in fact, litter the medical pages of the latter half of the nineteenth century. Sequin tells the sad story of a blind American slave named Tom Bethune who was auctioned with his mother in Georgia in 1850 for nothing because he was thought to be useless. But Blind Tom, as he came to be known, became one of the greatest musical savants of all time, able to memorize pieces at the piano by Liszt and Beethoven merely by listening to them a single time. In 1842, at age eleven, he played to President Buchanan in the White House, and he later toured Europe with his master, one Colonel Bethune. Tom's performances were apparently quite bizarre. After announcing that "Blind Tom will play this or that piece for you," Tom would sit at the piano and adopt a curious position, while playing "with an unknown force," as Sequin has it, "which manifestly proceeds from powers higher up than his wrist."

After regaling his audience with wild laughter, face-rubbing, and "some uncouth smiles," Blind Tom would then go into a kind of musical and acrobatic trance that cannot but remind one of the groaning and rocking of Glenn Gould, though in a far more exaggerated form. Sequin goes on:

> As soon as the new tune begins, Tom takes some ludicrous posture, expressive of listening, but soon lowering his body and raising on one leg, so that both are perfectly horizontal, and supported upon the other leg, representing the letter T, he moves upon that improvised axis like the pirouette dancer, but indefinitely. These long gyrations are interrupted by other spells of motionless listening, with or without change of posture, or persevered in and ornamented with spasmodic movements of the hands. . . .

That Blind Tom was a musical genius was taken for granted by his audiences, even if he was largely a kind of glorified circus attraction. (The journalist Edward Podolsky described Blind Tom applauding himself violently at the end of his performances, "kicking, pounding his hands together, turning always to his master for the approving pat on the head.") Yet in all other areas he seemed to be *ament*—that is, devoid of normal intelligence. "Most aments are fond of music," Tredgold observed, and Treffert adds: "The association of musical ability and mental retardation is frequent throughout the literature on the savant."

A hundred years after Blind Tom, in 1969, another American musical savant simply known as Harriet entered the literature, courtesy of the Boston psychiatrist Dr. David Viscott. Harriet possessed an IQ of about 73 and was admitted into Viscott's hospital depressed, psychotic, and hallucinating. Yet her musical intelligence approached genius level. She could recall thousands of compositions from memory and could play both the violin and piano. Not only that, Viscott writes, but also over the years she had memorized "the name, age, address, family structure, indiscretions, marital problems, and personal musical history" of every member of the entire Boston Symphony Orchestra. She could also give the date and weather for every day it had been given on the radio, had committed whole pages of the Boston phone book to memory, and could remember any event that had ever happened to her, though seemingly without any trace of feeling. One of her feats was to play "Happy Birthday to You" à la Mozart, improvising the relevant harmonies around the bass line of Mozart's first piano concerto to fit the birthday tune. She could also do this in the style of Prokofiev, Schubert, and Debussy, yet was unable to define even the simplest common words. For Harriet, Viscott concluded, "music became a special language and it remained the language of an endless infancy."

As the idea of genius has become increasingly discredited in the wider culture, it simultaneously has become increasingly medicalized. Talking to American graduate students, I often hear the casual remark, "There's no such thing as genius," often swiftly followed by the standard observation that, "Great men don't exist." It seems to be a mantra routinely handed down in liberal arts faculties across the country. And one can see why: it's a sentiment that appeals to American notions of equality. It is equality that is *normal*, not genius.

Even in the 1850s, Blind Tom may well have been more famous and popular in America than the Liszt he so uncannily imitated. A blind American-born slave boy, exciting both sentimental pity and scientific curiosity, may have been more accessible to the American mind than a haughty Hungarian domiciled in luxury in distant Paris. American democracy, indeed, would probably always choose a Blind Tom over a

Franz Liszt, for a Blind Tom doesn't threaten the sacred laws of democracy by being superior to everyone else—he's merely a scientific freak, like Rain Man.

★

But who was Rain Man? The fictional character of Raymond Babbitt in the film was based on an autistic savant by the name of Kim Peek, whose remarkable feats have been recorded by the film's writer Barry Morrow, who first met him in 1984. The thirty-three-year-old Peek astonished Morrow by reciting from memory the zip codes of every member of the Association for Retarded Citizens, as well as regurgitating endless amounts of baseball trivia. His popular nickname among his family and friends was Kimputer.

In his 1996 book *The Real Rain Man*, Kim's father, Fran, describes Peek as being a very Aspergerish autistic: "a warm, loving personality" who is also an accidental expert in a dizzying variety of largely useless subjects. These included the space program, the Bible, Mormon Church doctrine and history, telephone Area Codes, TV stations and their markets, Shakespeare, Kentucky Derby winners, details of 7600 books, and calendar calculations. "He could also," his father went on, "describe the highways that go to a person's small town, area code, and zip code, television stations available in the town, who the persons pay their telephone bill to, and describe any historical events that may have occurred in their area."

Peek's father is at pains to stress the usual lessons of American philanthropy. "You don't have to be handicapped to be different," he cries. "Everyone is different!" But the fact remains that the autistic savant is not a bright and merry mascot for the cult of Difference. Most bear no resemblance to the adorable Dustin Hoffman. Their difference is rarely approached in real terms, for it's obviously far more pleasant to marvel at their incredible feats from a distance. What lies behind the performing bear facade of a man like Jerry? I cannot help thinking that he must be a little like Edward Pullen.

Nevertheless, the genius has now been transformed into a lovable ament, a Forrest Gump with divine skills. Americans love what could be called divine idiots, a type brilliantly embodied by Harpo Marx. The whiff of divinity is what marks Rain Man out as a genius, because genius must still be magical in some way. Hoffman is said to have remarked rather mawkishly to Peek, "I may be the star, but you're the heaven!" But mawkish or not, Hoffman probably meant what he said, and we can more or less understand the attraction.

What's off-putting, though, is a little harder to describe, or perhaps just harder to admit to. I, for one, have even found a certain sinister quality to Harpo Marx. Especially as a child, I found him absolutely terrifying. I often had nightmares about him blowing bugles, with his buggy eyes and blond Afro hair. And Jerry has this same quality, which is hard to pin down intellectually. Perhaps it's a kind of knowing unpredictability; perhaps it's the zany gift for logical illogicality. Both Jerry and Harpo are men-children hovering between an uneasy childhood and a scathingly mocked adulthood.

★

When we got into Manhattan, I left Jerry and Mary in the psychotic glare of the new Times Square. For a moment I felt a twinge of guilt leaving them in such a devouring place, two innocents singing Herman's Hermits songs arm in arm. But after all they were adults, and Mary knew the address of the hotel. They were going to walk around Times Square first and take in the sights—giant faces of anchormen and basketball stars spread over screens hung in the heavens.

Jerry shook my hand vigorously and almost clinically through the car window:

"Good luck getting back to Brooklyn, Larry."

He winked at me as if we two now had a little secret, and as if he were concerned about me, not the other way around.

"See you at the Autistic Conference in San Diego. You know what we're going to do?" He still, I noticed, had the Willy the Whale soda cup firmly in his hand, and his eyes had become momentarily feverish, malicious. The Asperger anarchist suddenly burned bright, fueled by a lifetime's subtle resentments.

"We're going to burn the *Diagnostic Manual* on the beach! See ya!"

CHAPTER 5

DIAGNOSING JEFFERSON

The whole of my life has been a war with my natural tastes, feelings and wishes.
—Thomas Jefferson

It was already past dusk, and I was driving alone through downtown Kansas City in the last hour of pink prairie light. I was lost among emptied streets, rolling boulevards that rose and fell like fairground rides and in which only the bail bonds offices were alive with a greenish luminescence and a sequestered human hum: a city like an aquarium, its sidewalks swept clean of all citizens but the drunken nomads of its bar life. Somewhere among the Heritage Steak Houses and pylon spires of the Convention Center, I thought, the old Wild West town must still be lying like the compacted sediment of a vanished city. Perhaps those banks with ornamented clocks and barometers belonged to it, along with those iron facades and the Deramus Building lit up in gold like a ruin? But I couldn't say. An obsession of mine with American cities—a suspicion—is that more and more they are coming to resemble each other, and none of them naturally breathes the past. And which way was out?

There was no one on the street to ask at 9 P.M. One would have to venture into a steakhouse in a restored meatpacking warehouse or a bar of dubious custom and ask. So I crawled past neoclassical piles on West 11th Street while trying to decipher a metropolitan road map, for, in fact,

I was trying not to tour Kansas City but to leave it. I was destined for the remote suburb of Stilwell, Kansas, and it was only by accident that I had stumbled into Kansas City's aquarium-like downtown at all. Stilwell, I kept repeating to myself, I'm going to Stilwell and nowhere else.

My map, in fact, contained all the information I needed to get to Stilwell, but I was still unable to read it. The tangle of freeways knotting the intestines of the city was too tight to unravel without a magnifying glass. I could see Stilwell, some ways down to the south, but I could find no simple way of getting to it. There was a conundrum.

Therefore I coasted for an hour up and down little hills, along a Broadway of iron grills, past the Union Bank, the Venetian windows of the deserted Hotel Pleasant and a Gothic Holiday Inn, where the National Rifle Association had posted a large welcome sign next to that of the McClunary Wedding Reception. I was getting more and more lost as the evening wore on, and more and more unable to avoid passing the bail bonds offices, which were filled with large, gay crowds who, I thought, were beginning to eye my car with suspicion. Perhaps bail bonds offices are where people go at night, given that it's so difficult to get to Stilwell.

Finally I came to a hill more prosperous-looking than the others, with a view over the city and a monument to Thomas Jefferson. I stopped and got out to gather my wits. Was that really Jefferson standing tall above Kansas City in the summer night with an Indian and a leathered frontiersman flanking his shoulders? It must have been, because he certainly had Jefferson's nose and crazy hair. And besides, it said so on the pedestal. There was even a quote from Jefferson, which I unfortunately neglected to write down. There was no one there, except for two couples apparently making love in their cars. No one, certainly, had made the journey up here to gaze at Jefferson or, with him, over the great expanse to the west.

Jefferson calmed me down. The truth of the matter is that I have always liked Jefferson, and even the thought of Jefferson is enough to put a smile on my face. It's a complex phenomenon, but I shouldn't have to defend the fact that I like Jefferson. This is his country, after all. When it comes down to it, what I like about him is his complete absurdity: his protean absurdity, almost. Jefferson's was a creative, amiable absurdity, a capacity to be open toward many aspects of life. Apart from the obvious things, like authoring the Declaration of Independence and being the third president, we remember him now for his foibles and passions: his secret black mistress, Sally Hemmings; his inordinate love of French wine; his obsession with architecture. When I walked around the Maison Carré, the exquisite Roman temple in Nîmes, I could only think of Thomas Jefferson, who walked around it himself in an obsessive way and declared it to be the most perfect building in the world—a pronouncement that subsequently caused a thousand American buildings to assume the delicately mysterious form of the Maison Carré. He was an eccentric, but he had none of the boringness of most eccentrics, let alone the genuinely mad. Jefferson was queer, but his queerness came from an excess of focused energy—a boundless hope, so to speak. Besides, it was because of Thomas Jefferson that I had come to Kansas City in the first place.

I was going to Stilwell, because it is there that Norm Ledgin, author of *Diagnosing Jefferson*, lives. Norm Ledgin is the father of a teenaged Asperger boy and has written a book that he believes proves that Jefferson also suffered from this mysterious condition. Norm, in fact, is one of those amateur historians who are such a remarkable feature of the American cultural scene—nonprofessionals who often succeed in influencing the more serious history practiced inside universities. In a democracy, after all, every citizen-researcher with a computer is considered to be potentially as good as any other citizen-researcher with a computer. American history, in particular, has become something of a free-for-all in recent years, with a widespread grassroots dissatisfaction with academic history, making itself felt in thousands of self-help history books on everything under the sun, from the Indian Wars to the Underground Railroad to the meaning of the Fourth Amendment—and including, it would seem, an Asperger's diagnosis of Thomas Jefferson.

I pursued Route 69 in the dark toward Antioch. On either side of the freeway, one sees other freeways gradually converging in asymptotic curves with one's own, then veering away again into more darkness. Except that there is no real darkness here. The prairie has been turned into a plantation of malls—the largest malls I have ever seen. People often complain about malls, but you have not seen a mall until you have voyaged to Antioch on Route 69. These malls are like imperial stables: football-field parking lots—low buildings that seem almost to be on the horizon, and utterly empty most of the time. Have they been built simply for the fun of building them? Do they seem resoundingly normal to the people hurtling around them? But to get to the malls you need roads, and Stilwell-Antioch is reputed to have more roads per capita than anywhere on earth. I was soon to learn the bitter dimension of this sensational fact.

For I still could not find Stilwell. I was now floating past grandiose gated housing developments in the Old Englande or Prairie De Luxe style, with names like County Shire, Nottingham, and The Home-steade. In between them stand airy modern Baptist churches with stained glass windows, then yet more malls. Like the interior of the St. Louis Hilton, each intersection is indistinguishable from all the others, every interchange is interchangeable. When what you are seeing through your windshield does not vary, driving itself becomes neurotic, obsessive. You begin to look for the tiniest markers or landmarks that will give you some bearings. That tuft of grass . . . was it the same tuft of grass I saw fifteen minutes ago by the Walgreen's parking lot, or is it a completely different tuft of grass altogether? Is it northeast or due southwest from the rusted traffic sign I saw at the edge of the County Shire housing development? Is that the broken drain that points toward the Early Rise Bakery, or am I lost?

In these conditions driving itself becomes, frankly, a flatly Aspergerish activity. Outside of the safety of my car, I am lost; inside it, I am on sure ground. After all, I know all the dials, all the radio buttons, all the gadgets. I can go, more or less, where I want; at, more or less, whatever speed I want. The car is a bathysphere I control. But after a few minutes of circling malldom, I began to feel my own incipient madness coming gaily to the fore. If I am not mad, I thought against my will, then "it" must be. For does one stay normal in a landscape so poignantly abnormal? People frequently inveigh against the "tyranny of the car," but I must confess that I love driving, and that I love being in cars. Yet now I began to wish I was not scouring the roadside for cryptic signs and not looking at the illuminated dials of my dashboard for com-fort. I suddenly became aware that what I like about driving is com-pletely artificial: a sense of rapid change. Take the latter away and driv-ing becomes the torture that it always secretly was. An autistic, solitary torture.

I began to suspect that, far from taking me directly to Norm Ledgin's doorstep, Route 69 was implicated in a stupendous conver-gence with other, equally imposing road constructions, namely Metcalf

Avenue, 119th Street, and Blue Valley Parkway. How these constructions made sense of each other made no sense to me whatsoever, and
probably make no sense to anyone but the engineers who designed
them. And so, utterly bewildered, I ventured foolishly down 119th
Street, only to find that it was not a street at all but a rural highway
ploughing through miles of flat grass, underpasses, and yet more malls.
I realized that I now had to get off 119th Street as soon as possible,
before I was whisked away farther eastward into the depths of Missouri; and not only that, but the Ledgin family would be waiting for me
and wondering where I was. Immediate action was called for.

I now made the mistake of doing something spontaneous. I had
come to a desolate intersection with hanging lights rocking in the wind.
There was not a car or a person to be seen. I decided to make a left into
a bank parking lot and make a U-turn. But I started the maneuver suddenly and neglected to use my indicator. As soon as I had turned the
wheel leftward, a deafening siren exploded and a state trooper swung
behind me with his lights in full hysteria mode. A searchlight beamed
through the rear window. A cop on 119th Street in the middle of
nowhere? But on American roads today, it is perilous to look lost or
estranged in any way, to deviate from what locals know by heart. Our
roads are supervised to the most extraordinary degree, and the slightest
lapse from the Highway Code causes an instant reaction from the
legions of officers hiding behind every other fence and hedge.

A rather charming red-faced boy in jackboots came to the window.
He pointed out that I had not used my left indicator as I turned into
the parking lot. He looked half terrified and held on to his hat in the
wind. I put on my Prince Charles accent.

"Actually," I said, "I'm looking for Stilwell."

"Stilwell?"

He rubbed his red face. I could see that he was relenting.

"You're going the wrong way for Stilwell," he said. "Stilwell's the
other way. You'll have to make a turn at the bank here."

"Yes."

"To get to Stilwell you'll have to go back on 119 till you get to Blue
Valley Parkway, then hang a right onto the Parkway, then hang a left
lane to get to Route 69 without going as far as Metcalf Avenue, then
keep an eye out for the sign for Antioch. Clear?"

Then he asked me what I was doing in Stilwell. I replied that I was going there to investigate a thing called Asperger Syndrome. "You've probably never heard of it," I added.

The cop definitely relented now and handed me my license back.

"Sure I know what Asperger Syndrome is." He crouched by the window and holstered his flashlight. "I got a nephew has Asperger Syndrome. My nephew in Topeka. My brother got him diagnosed at the University of Kansas. You know all about the University of Kansas, don't you?"

I did not, but I quickly pointed out that they were world famous, brilliant pioneers, numero uno leaders in the field.

"That's right. We're the leaders in the field here in Kansas. And that's where my nephew got diagnosed. He perseverates."

"Perseverates?" I said. "Well, that's a symptom."

"Perseverating is actually the number one symptom, if I'm not mistaken."

"You're right. So he perseverates, huh?"

"Oh, gosh. But then so do I!"

"Don't we all?" I said facetiously.

"Well, I guess I have to be going. So long, sir. And do remember to wear your belt and use your indicators at all times."

"I will do so unfailingly, sir."

"OK, your Lordship."

He saluted and laughed, and I was in a good mood again. It was one of those conversations that can somehow only happen far from cities.

However, this did not change the fact that his instructions made no sense, and as soon as I tried following them I was lost again. I began to pass the same eerie churches, the same mock Tudor villages, the same malls. Then, turning in great bumbling circles, I stumbled fortuitously onto the tail end of Route 69 and Antioch, which, as far as I could see, was not a town at all but another sweeping road that disappeared into yet more despoiled prairie. I followed it and soon was sailing through what looked like farmland, with little apple trees bent against the wind.

★

At a genteel crossroads in Oxford Township, I found the Ledgin home —a one-story bungalow surrounded by a clean white fence. As I drove up, a dozen lamps came on automatically, there was a frenzied yapping of dogs, and an anxious-looking, middle-aged man came out onto the porch in slippers. Norm had a kindly, rather fussing face, and I liked him at once.

"So you got lost?" he said without the faintest trace of surprise. "Welcome to our horrendous freeway nightmare. They've massacred the prairie, you know."

"I'm sure they did," I said, shaking his hand.

It was now rather late, and I wondered if I was imposing. But Norm was clearly anxious to talk about Jefferson—and the freeways. In fact, after retiring from the newspaper business, Norm had been a Kansas State road inspector.

"When we bought this house," he said, as we lingered for a moment on the lawn, listening to the singing frogs, "it was prairie. Now the developers have buried every inch of it. Against our will, too. I wonder what Jefferson would have said about that? It's an outrage."

Shrugging, we went inside. The Ledgin family—Norm's wife Marsha, their seventeen-year-old Asperger son Alfred, and fifteen-year-old Nick—was assembled in comfy chairs watching an ice-skating tournament on the largest TV I have ever seen in my life. Between them sat four yapping dogs who seemed displeased by my presence. The family waved as Norm ushered me past, toward the kitchen. I noticed at once a grand piano covered with a pile of dolls and teddy bears, a few Herb Mignery–style cowboy sculptures, diminutive red lamps, brass cockerels on decorative shelves next to china cockerels, a Tune Time Bugs Bunny clock on one of the walls, and a little cactus garden in a glass case. "We'll talk," Norm was saying, "by ourselves for a while. Why don't we go down and look at the indoor swimming pool? Built it myself, you know."

As we wandered around the DIY pool house, Norm continued his peroration on the road situation. He'd worked for the Kansas division of the National Safety Council for nineteen years, conducting safety education for various communities. Traffic safety, safety belts, traffic enforcement, that sort of thing. He'd even won the NSC a few awards in his time. He was passionate about safety belts. He was also passionate about urban sprawl. The 1968 race riots, white flight, the desertification of the American downtown: it all led to one thing.

"Developers rule," he said grimly, peering at his own neat waterproof tile work. "You can't plan mass transport because of the urban sprawl. We're getting hemmed in here; you can't move. And it's all totally pointless. What are all these damned roads for anyway?"

"You've got me there," I said.

"It makes me feel that it all happened for the wrong reasons."

We went back upstairs to the deck and sat under some loud wind chimes. Norm sighed and looked across his smattering of blossoming trees. He looked a little like a kindly schoolteacher who has been mercilessly exasperated by his administrators.

"Roads," he sighed a second time. "What can you do about them?"

Then he suddenly changed the subject; he brightened up.

"I happen to have an interest in the legislative elections of 1857."

"The elections of 1857?"

"The 1857 elections in Oxford Township. It's one of my special areas of study. It was a duel, you see, between the free-staters and the pro-slavers."

Norm leaned back, he laid his head against his palms. The night air was humid with an approaching storm, one of those *Wizard of Oz* storms, I thought, that make the prairie so quietly menacing. According to Norm, the 1857 legislative elections in Oxford Township (that is, in this very neighborhood) had led directly to the Civil War. I quickly discovered that he was an encyclopedic fount of knowledge about the statistics of every Civil War-era vote in Oxford Township. As soon as he started on the subject, I could feel that he was going to get carried away —and so he did. The numbers came thick and fast, in astonishing displays of memorization. Here were the legislative combats of 1857 in all their numerical glory. I noticed that the wind was picking up and that a muffled flashing storm was taking shape on the horizon. The apple trees began to twitch. In mid-sentence, Norm looked up, blinked like the deer in the proverbial headlights.

"Well, I guess we'd better be going inside. I'll tell you about the elections later if you're interested."

We retreated into the house. From a hidden room came the sound of demented piano playing—snatches of Beethoven suddenly interposed with crashing heavy metal chords. The effect was vaguely Bram Stoker. "That's Fred playing," Norm said, winking with sluggish pride of which only fathers are capable. "He can switch from *Moonlight Sonata* to Nirvana in a twinkle."

I noticed now that there was a sort of honey theme to the home's decor. There were pictures of the Three Bears pawing pots of the stuff. In the bathroom the wallpaper had a bee theme, and there were small honey bears on racks, model bees sprouting from tiny springs, and a *Home Sweet Home* honey bear sign. "Bee shy," read a wallpaper ditty out in the hall, and "bee happy, bee mine." I began to wonder if there might not be a Freudian reason for this, or whether Norm actually kept bees out there in the garden. But before I could investigate these charming possibilities, it was time to be introduced to Fred, who had stopped his pounding on the keys and was hovering in the kitchen with a look of wondering curiosity. He reminded me at once of Data, the robot character in *Star Trek* who spends many an episode trying to figure out the bewildering emotional life of humans. Data looks like a human except for his strangely whitened eyes, but his imitations of human intuitions are always off-key, laborious, and clumsily studied. His friendships with humans are tragically counterfeit. What reminded me of Data in Fred was his manner.

He's a beautiful boy: with curly hair, pale eyes, dreamy looking—a pre-Raphaelite ephebe. Whenever I asked him a question, his eyes fluttered, he stammered, and then stuttered a soft-spoken reply as if seized by some subtle seizure. First, though, he ventured to ask me a question.

"Where will you be staying tonight? A motel?"

"Red Roof Inn," I said.

"Seventy-nine, ninety-nine plus tax," he shot back, his eyelashes batting furiously. "Special weekend rates, no swimming pool, laundry service, telephone number 734-5678, fax number 734-5600, satellite TV, buffet breakfast included."

"Why Red Roof Inn?" Norm said.

"I don't don't know. I always go to Red Roof Inn."

"What about Best Western? It's right on the highway. Isn't it, Fred?"

"Yes, that is correct."

"I don't don't know. . . ."

"Eight-five plus tax," Fred piped up. "Special weekend rates, swimming pool, laundry service, telephone number 734-4567, fax number 734-4500, satellite TV including HBO, buffet breakfast included."

"Then again," Norm said, "you could also go to the Motel 8 or the Hampton Inns, not to mention the Drury Inn. Fred?"

"Well," said Fred without missing a beat, "the Motel 8 is only fifty-nine ninety-nine including tax. No swimming pool, buffet breakfast, local calls free, telephone number. . . ."

"I think," I said, "I'll take the Red Roof Inn. The Red Roof Inn agrees with me."

Norm laughed. "The Red Roof Inn, eh? We can call them if you want."

"No, no. All I want to know is how to get there."

Fred's eyelashes were now batting yet more furiously. He stammered excitedly that he knew all about how to get to the Red Roof Inn.

"Fred," said his father proudly, "takes a great interest in the Kansas City road system. He knows it by heart, don't you, Fred? Tell him how to get to the Red Roof Inn."

Fred positively trembled. "If you like," he offered, "I could draw you one of my road maps."

"He's memorized every road in the state. Just like he's memorized the vital statistics of every motel in Kansas. And some in Missouri, too."

Now imbued with a mission, Fred blushed with on-rushing energy and prepared to go to his room to draw me a road map.

"It's true," he said before leaving. "I've studied the Kansas City road network very carefully, and I've decided that there's definitely some room for improvements. I've drawn up some ideas for the improvements. The feed-ins and the signs are not very well designed, the lane mergings are a mess, so I've come up with some alternatives. Would you like to see them?"

I said I'd be delighted to see his ideas for the reform of the traffic system.

"I have whole books full of them. I've been drawing them for years."

"Speaking as a traffic professional," Norm put in, "I'd say that they are outstanding contributions to our understanding of the Kansas City road system. Fred sees it all so clearly. He's got a talent for traffic, that's for sure."

"Do you like driving?" I asked Fred.

"I love driving. I like traffic problems. I like looking at traffic system layouts. I think Route 69 could definitely be improved. The right lane could be extended to the 135th Street exit, for example. I have quite a few ideas for extending lanes, actually. They should have double-lane exits. And the signs need to be black on yellow rectangular ones."

"See?" said Norm.

"I've redesigned quite a few traffic circles too."

Fred blinked shyly, then his eyes went wide open.

"They should give him a job," cried his father.

"I'd like that map," I said.

"OK," Fred stuttered, "I'll be back in a minute."

Norm took me into his study.

"You know," he said, "at first Fred was in denial about his condition. He just wouldn't accept that he was Asperger's. Then he read my book on Jefferson, and he had to finally admit that the glove fit the hand. But he resisted for a long time. I knew all along. I remember when he was a kid, he was always saying 'Sorry' for things he hadn't done. When you said 'Good morning' to him, he never replied. I read all about it, and I realized that he was perseverating." Norm nodded to himself. "It was perseveration, and it was staring me right in the face: Asperger's. But I call it a condition, not a disorder."

The study was crammed with curios, including several busts of Jefferson. I noticed a few traffic certificates on the wall. It was through Jefferson, Norm explained to me, that he had come to diagnose Fred:

"While all this was going on, I was reading Jefferson. And I was struck by Jefferson's love of corduroy trousers. Fred would refuse to wear pajamas. He'd say that he didn't like the elastic band in the pajama bottoms. It struck me as amazingly similar to Jefferson's refusal to wear abrasive fabrics of any kind. He liked soft, flexible materials like corduroy, just as Fred does. It's an Asperger trait, I found. For that matter, when I started to compare Jefferson and Fred, I found twenty Asperger traits that they had in common. Twenty!

"One of them was music. Jefferson played the violin avidly; and Fred was planning to major in music. Jefferson would talk your ear off on obscure subjects, so could Fred. Jefferson indulged in socially inappropriate behavior, so did Fred. If you said 'Good morning' to Fred, he'd just ignore you—even if you were standing right in front of him, he'd act as if you didn't exist."

"What socially inappropriate behavior did Jefferson indulge in?" I asked.

"Well, he went after his neighbor's wife, for one thing."

"But can you diagnose a dead person?"

"That's a good question. I say, why not? You have to wonder about Mozart. And what about Paul Robeson?"

"Paul Robeson?"

"He was pedantic as hell."

"You have to admit," I said, "that it's a tricky question."

"It is a tricky question. But there's no doubt about Jefferson. Jefferson was a classic case. Besides, he's a role model for all American Asperger kids who think they aren't normal. I can say to them, 'Take heart, was Jefferson normal?'"

"You have a point," I said.

Norm was now trying to make us coffee. The Ledgins have a standard filter-drip machine, but Norm approached it with a deeply bitter wariness. He obviously had had a history with the machine and could not make head or tails of it. "Damn thing," he kept muttering darkly, trying to slot the filter into its receptacle. I could see that he had not grasped the principle of properly seating the sachet of coffee in the filter but thought it best to let things take their course. And so Norm grappled heroically with the recalcitrant machine, as if it were some sort of disobedient groundhog, cursing and slamming it, until the glass jug slowly filled up with murky hot water.

"Look at that," he cried, with a look of wounded yet reasonable exasperation. "I don't understand these damn coffee machines. I've never been able to figure them out. It isn't me who makes the coffee in this house."

At this point Fred reappeared. He had drawn out several maps in pencil on sheets of a notebook.

"Here," he said. "I thought you might need different routes."

The maps were superbly drawn, with unfaltering, pure lines that had clearly been executed without instruments. One of them was entitled, "Directions to I-70 from Overland Park Red Roof Inn." Suddenly all the confusions I had experienced in the earlier part of the evening were clarified. Far more lucidly and simply than any of the road maps I had brought with me, Fred's diagrams made sense of the insane tangles created by the Kansas City developers. The disposition of Metcalf Avenue, for example, with relation to 103rd Street and the meandering course of Interstate 435, was immediately rendered intelligible. The Red

Roof Inn was neatly marked with a small x near the intersection of 107th Street and Barkley Street, one block north of the large intersection of Metcalf Avenue and Interstate 435. In a corner of the page, Fred had written with extreme neatness "Note: this map is not to scale." This helped. None of the maps, I suspected, were actually to scale, which led to irritating confusions. But Fred's map at least made everything abstractly simple.

It was now getting late, and it was agreed that I would come back the next day for lunch and a more detailed talk about Jefferson. I was all set to go off to the Red Roof Inn, armed as I was now with at least four different maps dealing explicitly with this single problem. Before I left, Norm handed me a copy of his book and warned me that ignoring Fred's maps would be a certain route to navigational damnation.

"Trust me. No one knows our roads better than Fred."

"I'll use the maps," I said.

But in reality I had come to think that the whole thing was a bit of an exaggeration. I only had to find where Metcalf Avenue joined Interstate 435, for heaven's sake. I had an idea in my head as to how to do this, and this idea was already assuming a luminous certainty. So I set out for the Red Roof Inn with fair hopes.

An hour later I was somewhere on 103rd Street, lost and exhausted, looking for a Barkley Street that did not exist. The freeway system, I now realized, was in reality a large Snakes-and-Ladders board on which only fiendish cunning and concentration were rewarded. I began to swear at the windshield and then at factories floating by in the night. Then I realized that I was passing familiar constructions. County Shire, Nottingham, The Homestead, the same Tudor mansions and prairie timber bungalows, the same modernist Baptist churches in which Norm had assured me people spoke in tongues on Sunday mornings. I passed a vast sign lit up like a satanic bonfire:

CITIBANK
The Bank of the Upwardly Normal

Then, before I knew it, I had been whisked back to Antioch Avenue and was heading again toward the Ledgin residence. Although it was now too late to ask them directions all over again, I parked gratefully by the white fences and turned on the car's interior light so that I could peruse Fred's hand-drawn maps. This time I would obey them to the letter. Half an hour later I was in bed at the Red Roof Inn on Barkley Street, reading the first pages of *Diagnosing Jefferson*, while that night's *Iron Chef* played on the TV. When I thought back on it, the infernal traffic system around Kansas City was not nearly as indecipherable as I had at first imagined it to be. All that was needed to unravel it conceptually was a bit of autistic perseveration and a willingness to let logic have its mad way.

★

> My Jefferson study reveals a brilliant and talented man who made love to women, tippled his wine, wore funny clothes, abhorred making speeches, fled to seclusion, brooded and wept, sidestepped disputes, waffled on race, obeyed dumb routines, believed fiction real, endured painful losses, and finally went broke.

Norm firmly puts Jefferson in a kind of Great Asperger Tradition, a sweeping gallery of afflicted geniuses that includes Van Gogh, Darwin, Gregor Mendel, Einstein, and, of course, Bill Gates. But what makes Jefferson Aspergerish?

To Norm's mind there are obvious first clues. "To some his body language appeared odd and awkward. He sang under his breath constantly. Often he looked disheveled, and he drank too much." His relations with women were tortuous, as is suggested by his extravagant vow of celibacy after his wife's death and then his long-concealed affair with the family slave, Sally Hemmings. But lest it seem that Norm is merely grabbing at biographical straws, he states at once that he has scrupulously compared Jefferson's symptoms with the criteria for Asperger Syndrome laid out in the *Diagnostic and Statistical Manual*. There, he quickly finds five matches for Jefferson. These include "a marked impairment in the use of nonverbal behaviors such as eye-to-eye gaze, facial expression, body postures and gestures to regulate social interaction" and "apparently inflexible adherence to specific, nonfunctional routines or rituals." Another symptom, the Manual states, is "lack of social or emotional reciprocity." (To be diagnosed with Asperger's, one only needs to be matched with four criteria.)

The *Diagnostic and Statistical Manual* does not tell us what "nonfunctional routines or rituals" are, but the criterion could easily, I imagine, apply to most sport fans glued to the TV on Superbowl Sunday. It could also, in Jefferson's case, apply to a passionate interest in wine. Most culture, for that matter, is a nonfunctional ritual. But Norm has other things in mind, for it is clear that, according to many eyewitnesses, Jefferson was quirky and that his quirks were sufficiently numerous to warrant our psychiatric suspicion. For one thing, he began each day by soaking his feet in icy water. "I offer the observation," Norm writes, "that temporary discomforts appear to serve Asperger-related needs for physical pressure, needs which the scientific community has not yet elaborated fully."

Jefferson's other peculiarities seem to have been abundant. He was said to have paced back and forth or in circles when distressed, to have spoken in a strangely "swallowed" tone of voice, to have exhibited ambidexterity, to have been a loner, to have sat awkwardly, to have neglected his grooming needs, to have been hypersensitive to selected dissonant sounds while being simultaneously hyposensitive (tolerant) toward certain other loud noises, to have been stiff in posture, to have worn an inexpressive face and a faraway expression, to have seemed unable to sense others' feelings, to have cultivated emotionally detached types such as engineers and scientists, to have suffered frequent headaches in youth, and to have had a propensity to diarrhea. All of these symptoms, Norm writes, Jefferson shared with Asperger Syndrome.

But perhaps the most Aspergerish of Jefferson's obsessions, Norm argues, was his lifelong tinkering with his unfinished mansion at Monticello. "Jefferson's dedication to the Monticello project," Norm writes, "met the *Diagnostic and Statistical Manual* criterion of encompassing preoccupation with one or more stereotyped and restricted patterns of interest and is abnormal either in intensity or focus." But more than that, he adds, "perhaps as many as a half-dozen separate Asperger traits as well went into the long-term construction of the dwelling." Monticello, the temple of Jeffersonian democracy, is in reality a temple to Asperger perseveration.

Monticello certainly obsessed Jefferson just as it also eventually bankrupted him. Indeed, in his book *Jefferson's Monticello*, William Howard Adams called it "the quintessential example of the autobiographical house." Secluded, impractically situated on top of a mountain, Monticello incarnated both Jefferson's yearning for withdrawal (it was originally to be called The Hermitage) and his equally fierce defiance of common sense. It thus represented a state of mind, according to Norm, "closely tied to Asperger's behavior."

The house was stuffed with technical subtleties and innovations. Apart from a clever use of half-octagon forms and mechanically operated skylights that admitted air but not rain, Jefferson took pains to devise dumbwaiters which were strategically placed so as to reduce the movement and visibility of servants. Because the latter were mostly slaves, and because Jefferson is deemed to have been what Norm calls "a reluctant slave owner," this proclivity for dumbwaiters, we are told, exhibits a strong sense of shame. Norm unhesitatingly credits this to the typically Aspergerish sense of right and wrong, which is always severe and uncompromising. He also praises Jefferson's "visual thinking," to which he attributes Monticello's original disposition of utility rooms, some of which were placed underground so as to better service the main house.

A remarkable thing about Monticello was that it was not built in one fell swoop. Instead, it festered and bubbled and morphed endlessly in Jefferson's mind, expanding in incremental additions, alterations, and slowly evolving inspirations. Fifty-four years after it was begun, it remained unfinished. Moreover, while this interminable construction was in progress, the house itself was barely habitable. Norm notes that this discomfiting of his family and visitors did not bother Jefferson, because "he was not a man of great empathy." At no time was Monticello actually comfortable to live in. It was a house-in-progress— a shambles.

Much of this had to do with Jefferson's childishness. His fickle love of mechanical devices was well known, as was his inclination to change plans abruptly *in medias res*. Grandin has said that Asperger Syndrome is like "a prolonged childhood," and that Jefferson's attitude toward the design of Monticello reminds her of herself: it was practical, logical, and picture-based rather than verbal or emotional. Why else, asks Norm, would he have flattened the posts of the stairwell balustrade purely for the sake of aesthetic effect?

As soon as Jefferson is suspected of having a syndrome, all of his behavior becomes suspect. His habitual and obsessive note-taking, for example, or his eclectic interests in obscure subjects. To be this way now seems abnormal. In his "Letter to Lord Byron," W. H. Auden described this taste for pedantry and connoisseurship fondly, if not nostalgically:

> To be a highbrow is the natural state
> To have a special interest of one's own,
> Rock gardens, marrows, pigeons, silver plate,
> Collecting butterflies or bits of stone;
> And then to have a circle where one's known
> Of hobbyists and rivals to discuss
> With expert knowledge what appeals to us. . . .

That is Jefferson's temperament in a nutshell. Similarly, Jefferson's famous avoidance of conflict and brutal truthfulness, his tendency to tell different people different things in order not to displease them, was not necessarily a sign of some autistic ineptitude. It was more likely what Joseph Ellis in *American Sphinx* called a "diplomacy of the interior regions." Courtesy was an art for him; it took precedence over uncompromised sincerity. "In other cases," Ellis wrote, "it was an orchestration of his internal voices, to avoid conflict with himself." These inner voices, Ellis claimed, often simply couldn't hear each other. So that one minute Jefferson could write a letter expressing his adoration for all things French, and then, a day later, he could write another giving forceful vent to a garrulous gallophobia. Different ears, different songs.

Others took this as simple lying. In his essay "Amistad," Gore Vidal notes that John Adams, for one, found Jefferson to be a morally exhausting experience:

> He found dining with Jefferson exasperating, because of what Adams called his "prodigies," a polite word for lies of Munchausen splendor. When Jefferson was in France, he claimed that the thermometer remained below zero for six weeks; he then shyly confided that on a trip to Europe he had taught himself Spanish. "He knows better," J. Q. A. groaned, "but he wants to excite wonder." And admiration, something no Puritan Adams could ever do.

Lying, however, sometimes took a more dramatic form.

Norm Ledgin makes much of a curious footnote to Jefferson's writing of the Declaration of Independence. An early version of the legendary document known as Summary View, written two years earlier, traces the roots of America's debate with the British crown back to the

Norman invasion of England in 1066. Jefferson followed standard Whig history in claiming that the ancient Germanic democratic rights enjoyed by the Saxons had been crushed by the French-speaking feudal knights of William the Conqueror. It was the Norman Conquest, according to Jefferson, that inaugurated "the fictitious principle that all lands belong originally to the king. . . ."

Jefferson was much in love with the idea of freedom-loving tribes living happily in the forests. For him, ancient Germans and Saxons and contemporary Hurons and Iroquois all signified much the same thing. The noble savage was, of course, a popular conceit of the Enlightenment. But in the case of the Saxons, Jefferson spun his conceit more elaborately and then placed it at the heart of the Summary View. In the eyes of most commentators since, this was an unforgivably frivolous example of "fictitious principle"—i. e., lies of Munchausen splendor. Norm therefore sees it as an example of Asperger extravagance. Even Ellis admits that it's a "crucial clue to Jefferson's deepest intellectual instincts." The Saxon myth, in other words, was not an example of Jefferson indulging in a little harmless exaggeration. It was, as Ellis thunders, a "complete fabrication." Ellis is clearly scandalized by Jefferson's illusions, as he calls them, adding irritably that they "possess a sentimental and almost juvenile character that strains credulity." He goes on:

> The explanation lies buried in the inner folds of Jefferson's personality, beyond the reach of traditional historical methods and canons of evidence. What we can discern is a reclusive pattern of behavior with distinctive psychological implications. . . . Monticello offers the most graphic illustration of Jefferson's need to withdraw from the rest of the world, filled as it was with human conflicts and coercions. . . .

Jefferson's sentimental attachment to the Saxon myth, then, is as ludicrous as his obsession with Monticello, that "magical mystery tour of architectural legerdemain." The fussy tinkering with Monticello was "childhood play adapted to an adult world." The house it self and the his expectations for it "suggested a level of indulged sentimentality that one normally associates with an adolescent."

Norm, for his part, draws much the same conclusion. "Oh, he was a sentimental all right. A kind of child-man."

What is interesting here is how much psychiatry and history agree on Jefferson. It strikes me that Ellis's reaction is overwrought. Is Jefferson's belief in the Saxon myth really a sign of psychological disturbance, a reclusive pattern of behavior with psychological implications? Jefferson, it seems to me, merely had a utopianist cast of mind, as did many intellectuals of the eighteenth century, not to mention our own. Furthermore, isn't there a grain of truth in Whig history and its feeling that primitive tribes enjoy a greater autonomy of life than feudal serfs? It's a sentimental view of the past, but hardly a Munchausen-like lie. It is, rather, part of Jefferson's exuberance and generosity of spirit—qualities that are all too frequently allied to a voracious gullibility.

"In Jefferson's case," Temple Grandin writes, "specific-to-general thinking is what made Monticello such a masterpiece of architecture. Ask anyone to draw a floor plan for a house, and that person will start with a square or rectangle. Not Jefferson."

Thus Jefferson did not think like a normal person. He thought in pictures, just like Temple Grandin and like Fred. I looked at Fred's traffic plans one more time, before dropping off to sleep, and had to admit that it was easy enough to imagine a busybody, eternally improving Jefferson interesting himself in the laws of traffic in exactly the same way. Fred's drawings were pure pictorial thinking. They were like images of the American city seen by a passing bird.

★

I found that I was now anxious to talk to Norm about Jefferson and especially abut the latter's love life. Hadn't it been something of a fiasco? What about his relations with women? Norm's eyes lit up, and I thought I detected a faint glimmer of gleeful malice. Jefferson . . . women.

"Oh, Jefferson had nothing to say about his mother," he said, as we sat down in the living room and the giant TV blared away. "He was a tyrant with his daughters, too. He made bad choices for the women he proposed to and even the women he fell in love with. I mean, Maria Cosway wasn't a great choice."

Cosway was the flirtatious Anglo-Italian Jefferson fell in love with in Paris. He had broken his leg trying to jump over a wall in her presence, and the affair had petered out miserably.

"I suppose that's true," I said. "Is poor choice in women an Asperger trait?"

"I guess it is."

The family was gathered in front of the TV as before, and Marsha, Norm's wife, promptly piped up to the effect that Norm was as guilty of being Aspergerish as Jefferson himself. At first I thought she was joking, but her voice was deadly serious.

"You are, Norm," she rattled at him, "you are."

"Well," Norm murmured, "there are some areas—"

"For one thing, Mr. Osborne, he perseverates. You do, Norm, you perseverate."

"Yeah, I do." Norm shrugged with infinite resignation before the onslaught. "I do."

Marsha suddenly became quite animated in her comfy chair, as if she needed to get something out.

"He's fixated. He was fixated when he worked at the newspaper. He's fixated on semantics."

"Yeah," from a sad Norm.

"He's really quirky. When he's fixated on semantics, he'll get on some frigging toot."

"Yeah."

"He'll get on a toot. He gets on a routine, and then he's on a toot. Aren't you, Norm?"

"I do get on a toot," Norm admitted. "About words."

"He checks his e-mail at 4 A.M. He has to have six cups of chicory coffee. He has to have breakfast in the morning."

"Is that true?" I asked Norm. "About the breakfast, I mean."

I winked, but no one saw it. "It's true," was all Norm said.

"That's a rigidity," Marsha cooed triumphantly, wagging her finger.

"I don't don't always have breakfast."

"Oh you do, you do. It's a rigidity."

"Well, I'm a perfectionist. I'll edit till death. I love my dictionary!"

But Marsha was already on her warpath. "He can't read people," she went on quickly. "He can't read between the lines, he can't communicate. He doesn't get the nuances."

"I've got my nose in a dictionary all the time!"

"Norm, you're Asperger if I ever saw it. You misread people most all the time. Other people notice it. They think you're kind of nutty."

"My sarcasm is intense," Norm groaned.

"You belittle people. You belittle them because of the way they talk." And now an anger, a contempt crept into her voice. "And you're inflexible. That's a characteristic of Asperger's, isn't it? Inflexibility?"

Norm sighed meekly. "I'm an either/or person," he admitted. "I have some Asperger characteristics, I'll admit, but not enough to be diagnosed. I'd have to draw the line at diagnosis, honey."

"But you definitely perseverate."

"That I do, yes."

I noticed that the two boys were smiling at this dialogue. Nick rocked back and forth and guffawed. Marsha then told me that Nick been diagnosed with Attention Deficit Disorder at the age of two and that he'd been taking Dexedrine since he was five.

"Not only," Nick burst in impatiently. "Clonadine, Ritalin, Risperdal, too."

"That's right, Nicky. He's ADH bipolar, you see. Risperdal is the antipsychotic."

Gradually Jefferson seemed to recede into the background as the Ledgin family set about discussing its medical drug regime, from which only the stoic Norm seemed to be exempt. For the first time, meanwhile, I noticed that a thick handbook of medical drugs lay conspicuously on the coffee table. I thought of the Paxil ads that are now common on prime-time TV, an air hostess voice intoning, "You can visit someone you haven't seen in a while—yourself." Laughing faces dance around the screen as the same voice continues to list undesirable side effects ranging from constipation to sexual dysfunction, until a woman steps up to a smiling man and says, "Hey, I remember you!" Paxil, one thinks—*pax*, peace? Marsha explained that she had taken Prozac for fifteen years, and before that amitriptyline.

"I've been in clinical depression since high school. So it's thirty years that I've been taking medications. Now Norm, he's a bit OCD, but he seems to live with it OK."

I asked Fred if he was taking anything.

"Me?" His anemone eyes opened wide and he stuttered. "I'm still taking BuSpar. It's for my anxiety."

Norm now turned brightly to me.

"Say, do you have any relatives diagnosed with autism?"

Not as far as I knew, I said.

"You should look into it. I suspect my father of having had it. He could memorize sixteen-digit credit card numbers. You wouldn't call that a normal ability, would you?"

"I dare say not."

"To me, there's an autism spectrum that exists that's way beyond autism proper. Most of us, I expect, are on that spectrum somewhere. The spectrum is like a rainbow of variety, of different types. It could encompass much of the population. Do I have Asperger Syndrome? Sure! Don't you?"

Fred now stood up abruptly. He shook his sleeves slightly and walked around to the piano.

"Of course," Norm whispered to me, "it's 12:10."

Without saying a word, Fred sat gingerly at the piano and flexed his fingers. The dogs were now tussling furiously under the piano, yapping in maniac voices.

"Fred always plays the piano at 12:10 sharp."

In his heavy glasses and shorts, Fred looked like a seven-year-old boy who has suddenly grown to the size of an adult by drinking a magic potion. As he began to play, his form slackened, he began to sway a little, his eyes fixed on a book of music spread on the piano. I went round to take a look. It was a collection of Nirvana songs, which was certainly not what was emanating from the piano—the first ethereal strains of Debussy's *La Cathédrale Engloutié.*

"'The Sunken Cathedral'!" Norm winked at me. "He knows that one by heart."

Fred rolled through the luscious chord progressions effortlessly, filling the Ledgin home with the far-off thoughts of Debussy. I was utterly amazed. Through the windows, the last scraps of uncemented prairie seemed to freeze while Fred pedaled in bare feet, his eyes transfixed. My eye strayed to the bookshelves, where gold-tooled editions of *The Grapes of Wrath* and *Our Town* lay slanted without a trace of dust. "The Sunken Cathedral" rolled on, as oblivious to them as it was to the tussling dogs under the piano and the pictures of the Three Bears and their honey pots on the walls. Marsha and Nicky were still watching TV. Norm, who had presumably heard "The Sunken Cathedral" a hundred times, merely tapped his foot and smiled.

"'The Sunken Cathedral'," was all he could say, "is almost as tricky as 'Elite Syncopations'."

When Fred had finished, I applauded sincerely. It had been a superb performance.

Then Fred said, "Want to hear one of my compositions? I'll play you something I wrote when I was eleven."

Suddenly Fred's hands went into a manic shudder and his left fingers rolled over the keys to stir up a demonic bass, thunderous chords. It sounded vaguely Mussorgskian. Somber and tragic, this odd piece shattered the mood of the Debussy. I began to feel subtly alarmed. Fred had gone quite pale, his neck stiffened and his eyes almost bulging.

"I call it *The Dark Side*," he called over his shoulder.

It was only as I was leaving that I got to talk to Fred a little by himself. He had left school already (too easy) and was taking classes at a local technical college—in music, principally. His ambition was to get into Kansas State University. Meanwhile, he was enjoying learning van Beethoven. Did I know the third movement of *Moonlight Sonata*?

"It's loud, furious, rocklike. I like it for the same reason I like playing Nirvana."

I said I couldn't quite imagine Nirvana on the piano.

"No, no, it's amazing, very satisfying. I like it better than 'Elite Syncopations'."

Fred blinked, and looked utterly earnest.

"Though not as good as Rachmaninoff, of course."

I felt like asking him if he admired Glenn Gould, but then it struck me that such a question was too laboriously obvious. But Fred preempted me.

"Have you ever watched *Mary Poppins*?" he asked.

"Four times," I said.

"Is that all? I used to watch it all the time. I've seen it hundreds of times. I suppose that's perseveration, right?"

And now, behind the thick lenses, Fred's eyes came alive with a perfidious little glimmer of irony.

"Some would call it that," I said.

We wandered out onto the lawn with Norm. It was now hot, and from an indeterminate distance came the sawing of buried cicadas. The mowed lawn crackled under our feet.

"Have you got your maps?" Norm asked me.

"To get to Interstate 70," Fred began at once, "from the Red Roof Inn, you take 107th Street north to 105th Street, where you take a left onto Metcalf Avenue, and from there a left onto East I-435, after which you continue east to North I-435, and then exit to East I-70. Then you continue."

Norm laughed. "Amazing, isn't it?"

I thanked Fred and said I was sorry that I didn't have anything as pretty as his maps to give him in return. But he shook his head.

"Don't worry about it. I have hundreds more. Probably thousands."

Saying which, he handed me a typed sheet that bore a short essay he had written about himself for presentation at a conference.

As I sat in the car, I heard more piano music wafting from inside the bungalow. Fred was just getting warmed up, hitting the keyboard with renewed ferocity, and the dogs had begun to bark with alarm. But I had to admit to myself that, at least on the piano, Nirvana didn't sound nearly as bad as Nirvana.

Fred had ended his essay with the following words:

> I could discuss my life much further if I had the time. But right now, I just want to say that my life has been interesting, and at this point, successful. Perhaps because it has not been normal.

CHAPTER 6

AUTIBIOGRAPHIES

Dear Hubert—
Get some more of my poems up on your Web site,
as this might put me in touch with a nice woman.
There's loads of women out there on the Internet.
                    —David Miedzianik

Camden Avenue must be one of the quietest streets in Los Angeles. The jacarandas seem to have been growing here for centuries, their roots spreading outward into the roadway, with gnarled knuckles and their fluffy lavender flowers falling onto the tarmac in a soft rain that reminds me of a provincial square in Mexico. The defunct trolley lines, too, seem to be the skeleton of a pre-city, making their way along long-forgotten streets. Yet a hundred yards away from the isle of jacarandas and the soporific lawns ringed with yuccas, Santa Monica Boulevard goes about its motions and transactions, and in the narrow walkways between the apartment blocks, there is the inevitable heaving of heavy metal music falling out of windows and of vacuum-cleaner attachments moving along glass slats. Halfway down Camden Avenue, one of these walkways is narrower and darker than most of the others. It ploughs claustrophobically between stucco walls and fences, blocked from the sun like a steep alley in old Genoa.

I was here to see the author of *Calculated Risk* and *Mr. Twiddle*, that same Jonathan Mitchell whose work is a legendary part of the Autistic Underground. I was a little nervous because I had almost no idea what to expect. Would Mr. Mitchell be a terrifyingly eccentric Asperger intellectual or a socially inept mass murderer with a slight proclivity for writing artistic short stories? I pressed the bell and waited, while cocking an ear toward the stirrings in the interiors of a dozen apartments all around. Everything was hideously muffled. Was someone blending carrots? Shrieking at a disobedient boyfriend? No one came to the door where I'd just rung, and no sound whatsoever stirred from inside, not even that of approaching feet. I tried again and again. Nothing. I found that I was sweating in the humidity—it was an oppressive day. I stepped back and looked up. All the windows of the two-story condo were fiercely curtained, and no view could be had of the ground floor. Perhaps he was writing? Perhaps he had forgotten the appointment altogether? For some reason I could not cast from my mind the possibility that I was facing the machinations of an unfathomable cunning. Perhaps, despite my enlightening exposures to the ways of autistics, I was still suspicious of them, a little fearful of a mind that does not function like your own, a sensibility with different laws. It reminded me of the feelings that assailed me in the souk of a North African city, trying by myself to find the keys to gates and passageways by reading a script I can barely decipher and by manipulating a language in which I can barely ask for a cup of coffee. So far, no unpredictable behavior had come my way, but that was no reason such behavior should not commence now. When Jonathan did not come to the door, I therefore began to feel uneasy. I decided to call up: "Mr. Mitchell?"

Almost instantaneously a nervous high-pitched voice barked back from behind the shrouded upper-floor window. "Who's there?"

Commotion, expressions of regret that the front door bell did not work, a tumbling of feet down a flight of stairs, fumblings at the door lock. I felt that Jonathan's eye was inspecting me from some secret device that was able to assess the character of strangers by means of an infrared beam.

"Is that Mr. Osborne, the reporter?"

I rarely hear that word used in respect to myself, and so I laughed. Why not?

"Yes, it's the reporter."

"You're six minutes late."

"But—" The door snapped open. Jonathan peered out: a stocky, richly tanned, unsmiling but flatteringly friendly man of about forty in his socks. He tried nevertheless to smile, as if such a gesture had been practiced for hours in front of a mirror. I noticed at once that his speech was stop and start, jolting. Inside, the apartment was plunged in near darkness, and Jonathan blinked in the light, mole-like.

"I've been waiting for you," he stammered. "Come in!"

It was a nice apartment, clean, well appointed. There was clearly money in the picture somewhere. I sat, and Jonathan flitted back and forth, darting into an office nearby to get some of his short stories for my perusal. He talked as he walked, and he walked in exactly the same way that he talked, which is to say in stops and starts, like a mechanical toy that is continually breaking down and starting up again of its own accord.

"Have you read Philip K. Dick?" he asked excitedly. "The science fiction writer? He wrote about autistics. *Martian Time Slip*. What about *The Sound and the Fury*? Well, everyone's read that one of course. I've been writing for seven years myself. I have 75,000 words of a novel. It's about an abusive special education school. It's called *The School of Hard Knocks*."

"Is it autobiographical?"

"Pretty much."

Jonathan sat down and calmed himself a little.

"Actually, we have a word for autobiographies written by autistics. We call them *autibiographies*. Cute, huh?"

"I've been told," I said, "that writing short stories is easier than writing a novel."

"For me it is. I can concentrate and get them finished. Novels are too elaborate. Sugar?"

He got up and bounded into the kitchen where the kettle was boiling. For a moment I looked down at the stack of short story manuscript pages he had deposited in my lap and scanned the first lines:

> *The Session*
> Andy Horowitz was nine years old and extremely hyperactive. In fact, he had been diagnosed as having Attention Deficit Disorder. . . .

I looked around the room again. There was no clutter, no sign of disorder, and above all no fatal sign of boxes. "I was in psychoanalysis for ten years," Jonathan called from the kitchen. "It didn't help."

At first I wanted to ask him why his house was in complete darkness, but after a while my eyes adjusted and I recalled that Asperger people often suffer from hypersensitivity to sensory stimuli, and that such stimuli must include light. Instead, I asked him about his writing, which promised to be a more interesting subject. But Jonathan, nevertheless, wanted to describe a few things about his life, as if he needed to be disburdened. In the first place, he had never been diagnosed; he placed little faith in diagnosis. He had never had a girlfriend; his loud voice usually put women off. Being from a family of Russian Jews, he disliked German Jews—wasn't Bruno Bettelheim one of them, a nasty piece of work? He went to UCLA, majored in psychology. He had worked in a warehouse, then at a clerical job in an insurance company, then had typed prescription labels at a pharmacy. He had worked at home doing medical transcriptions. He was always getting fired, he explained, because of his bad handwriting and his BO. He tried Prozac, but it didn't help. He had no friends, except for his schizophrenic pal Allen, who was also a writer. Jonathan had met him at Emotional Health Anonymous. "But then he had a breakdown," Jonathan sighed. "I lost him."

Jonathan had other writer friends, however. One of them was David Miedzianik, the Asperger poet living in the British town of Rotherham in Yorkshire, whom I had already discovered for myself. Jonathan had made one of the few voyages of his life to Rotherham to visit this fellow bard.

"How was Rotherham?" I asked. I knew Rotherham well.

"Rotherham? Rotherham was nice, but not that interesting. I don't know what to think about Rotherham. But I loved London. I loved the transport system. I wish we had a transport system like that in Los Angeles. I loved the Tube, the *A–Z* urban maps. Those *A–Z* books are really something."

I told him that I had grown up with those page-by-page geographical dictionaries of Greater London. They made fascinating bedside reading for future cabbies.

"Well, I devoured them," Jonathan said flatly.

"I don't have much to say about Rotherham either," I admitted.

"Of course there's no Tube in Rotherham, and no *A–Z* either."

I was almost relieved to find that Jonathan had all the usual writer's instincts, including bitter resentment at, and dislike of, almost all other writers, who were divided unambiguously and emphatically into friends and foes. The Englishwoman Donna Williams is probably the most famous autistic novelist, with her 1992 bestseller *Nobody Nowhere*. But Jonathan was scathing.

"It's a poorly written book. No chapters. I'd never write a book that had no chapter breaks!"

Then there was David Spicer, an Asperger poet in North Carolina.

"Oh, I don't like Dave Spicer. He's against my kind of medicinal regime. He can't tolerate any diverging views!"

And lastly there was the great Temple Grandin herself.

"I'm offended by Temple Grandin. I don't have a visual imagination. Please, that trivializes my suffering." Jonathan's voice grew angry, his hands became agitated. Was he flushing a little red? "She blows her own horn all the time. She brags. She makes generalizations. She brags. Let's just say that it's a mixed bag." He suddenly stood up and waved his hands around. "As for Bill Gates having Asperger's, that's ridiculous!" Then he sat down again, and we talked about celibacy.

"Well," Jonathan said, "Dave in Rotherham talks a lot about that. He has this poem called 'Next to No Feedback.' It has a refrain that goes like this:"

> Well there's next to no feedback on anything I do.
> To get some feedback from a nice lass would make me feel new.

"Dave in Rotherham talks a lot about that. I can sympathize. I feel the same way myself."

I read a few of his stories while Jonathan fussed around in the kitchen trying, I suppose, to figure out the mechanics of coffee-making. When he returned I remarked that there certainly did seem to be a fair amount of sex in his tales. There was a story called "Blot," which ended thus:

> Biff then penetrated Monique with his penis and all thoughts of Graham were put out of her mind. He was just a blot.

"Poor Graham," I said.

"Yeah. The autistic guy always gets shafted."

Jonathan had only had one date in his life, an outing in his late twenties with a fellow student in his Introduction to Fiction course. He took her out for coffee, invited her to dinner, and then she stopped returning his e-mails.

"These days," he sighed, "let's just say there's a lot of masturbating. Actually, I wrote a story about masturbating. I think you have it there in your lap. It's called 'Questionmark Etiology'."

I read it through. It was really rather droll, and I laughed.

"Pure autibiography," Jonathan grimaced. "Pure."

> Arnold Springer goes to a doctor when he feels excruciating pain in his hands. Sitting in the waiting room in agony, he rails against the new HMO's which oblige people to wait hours for a simple appointment. When the doctor appears, he is a disagreeably gruff Indian by the name of Dr. Blasubaramian. Dr. Blasubaramian gives Arnold's hands a thorough examination by means of an electromyograph machine, which records things such as nerve conduction velocities and the amplitude of the median, ulnar and radial nerves in the hands. Using things called Phanel's and Tinel's tests and the aforementioned electromyograph machine, the gruff doctor is able to diagnose Arnold as having a distal latency of 4.7 millimeters in the radial nerves, which indicates a clear-cut case of bilateral carpal tunnel syndrome.

Jonathan's prose style is remarkably matter of fact. He goes into the medical details of his tale with exhausting exactitude. As the story progresses, you have little idea what is happening outside these precisely related medical snippets. Why does Arnold have creaky wrists? We don't know. Arnold gets his diagnosis and goes home despondently, carrying a pamphlet about bilateral carpal tunnel syndrome and a plan to wear wrist splints and take steroid injections as part of a "conservative management" program. The good doctor, meanwhile, dictates into his Dictaphone the diagnosis, but adds that since he is unable to divine the cause of the carpal tunnel syndrome, he must classify it as a "questionmark etiology."

Once home, Arnold immediately grabs his penis. Alas, first one hand then the other proves too painful for the arduous task. The language suddenly becomes steamy:

Other guys could do well with the chicks and get their dicks wet, but not Arnold . . . he wondered what Blasubaramian would have thought if he had told him what he was doing with his hands and how he came to develop carpal tunnel syndrome.

Finally, completely naked now, Arnold decided to lay prone on the bed, rubbing back and forth on his stomach and trying to self-stimulate his penis without the use of his hands.

But he then remembers that it's Wednesday night and time for his weekly twelve-step Sex Addicts Anonymous meeting. On the door of the meeting room, we see this sign:

> Who you see here.
> What you see here.
> When you leave here.
> Let it stay here.

With this stunning pornographic flourish, the story ends. I put it down and started laughing.

"You hate it?" Jonathan suddenly looked alarmed.

"Not at all. It's very funny."

He looked even more alarmed. "Funny?"

"It's the contrast in tone."

My host looked a little darkly down between his feet and nodded. "I see what you mean," he muttered. "Funny."

Now Jonathan also reminded me of Jimmy Two Times.

"Don't you find it funny?" I said.

"I don't know. Sure. I mean, it's a sad subject, isn't it? But sometimes," he came back sharply, "it calms you down."

"Like writing?"

"Like writing. And like twiddling."

"Like what?"

"Twiddling." He looked at me defiantly. "Twiddling is one of my autistic things. I do it to calm me down. I twiddle frantically for a while, and I fantasize about all kinds of things while I twiddle. It helps quite a bit. Would you like to see me twiddle?"

I wasn't sure what to answer to this. The most nightmarish possibility was that twiddling was in some earnest way connected to masturbating. On the other hand, it could just as easily have something to do with writing. I was undecided. How exactly, I asked tentatively, did Jonathan go about twiddling?

"Well," he replied at once, "I usually use shoelaces tied around my fingers."

Although this was extraordinarily ambiguous, I decided to give it a go anyway. Jonathan charged upstairs and I was left for a time alone on the sofa with my cup of coffee. I heard muffled bangs and groans, and drawers being opened and closed. Once again, my unease returned. What had I gotten myself into? What could twiddling be but some deeply embarrassing or even menacing pastime with sexual dimensions? Nevertheless I waited quietly and kept an eye out the front door in case a swift exit became necessary. Eventually Jonathan reappeared. Two shoelaces dangled from each of his hands, and from one of them in turn dangled a pencil. He seemed geared up for a performance; I did not let the door out of my view. But Jonathan merely stood in the middle of the living room, then began rocking himself back and forth while twirling the two shoelaces around. After a minute of this, he began to jump up and down like a Masai warrior, the laces flapping round and round and the pencil with them. "I'm fantasizing," he cried, "about making a lot of money on Wall Street!"

"Is that twiddling?" I asked.

"Yes. Bruno Bettelheim described it in his book *The Empty Fortress*. It's autistic self-behavior, you know."

"I see."

"Of course, the neighbors complain. Things have gotten pretty bad between us. It's not as quiet as writing."

Thus Jonathan jumped, rocked, and twiddled, and I became mesmerized by his twiddling shoelaces and his rotating pencil. I hardly dared ask him what he was fantasizing about now. So I sat back and let him twiddle. It looked strenuous, vaguely masturbatory in its rhythms but, at the same time, aerobically heroic. Perhaps ideas for stories came to Jonathan in this way. Perhaps it calmed the diabolic inner forces of his solitude. I couldn't say.

"Well that's it," he said at last, coming to a standstill. "That's twiddling for you. It's part of the creative process. Without it I'd be lost."

★

If Hans Asperger was right that autistic thought is distinctive, there ought to be an equally distinctive Asperger writing. Jonathan's stories were written with a kind of solipsistic detachment that strikes an odd tone, both wooden and agonizing at the same time. In a school diary, Temple Grandin once wrote that "one should not always be a watcher—the cold impersonal observer—but instead should participate." In *Thinking in Pictures*, she comments further on this inability to live spontaneously inside images:

> Even today, my thinking is from the vantage point of an observer. I did not realize that this was different until two years ago, when I took a test in which a piece of classical music evoked vivid images in my imagination. My images were similar to other people's, but I always imagined them as an observer. Most people see themselves participating in their images. For instance, one musical passage evoked the image of a boat floating on a sparkling sea. My imagery was like a postcard photograph, whereas most other people imagined themselves on the boat.

Grandin theorizes that autistic children often use words in a purely associative way. For example, if a child associates the word "dog" with being outside, he will use the word "dog" to mean "outside." When she was six, Grandin learned to use the word "prosecution," which she associated with her kite hitting the ground even though she had no idea what it actually meant. "Prosecution" meant "downward spiraling kite."

Similarly, Grandin suggests that certain autistic minds navigate their way through the total disorder of a world they cannot grasp by means of simple, concrete images, which are arrived at fortuitously. For her, one such image was that of a door.

> In order to deal with a major change such as leaving high school, I needed a way to rehearse it, acting out each phase in my life by walking through an actual door, window, or gate. When I was graduating from high school, I would go and sit on the roof of my dormitory and look up at the stars and think about how I would cope with leaving. It was there I discovered a little door that led to a bigger roof while my dormitory was being remodeled. While I was still living in this old New England house, a much larger building was being constructed over it. One day the carpenters tore out a section of the old roof next to my room. When I walked out, I was now able to look

up into the partially finished new building. High on one side was a small wooden door that led to the new roof. The building was changing, and it was now time for me to change too. I could relate to that. I had found the symbolic key.

Thereafter, Grandin saw everything in her life in terms of doors. While preparing for graduation from college, she discovered a small metal trapdoor leading out from the flat roof of the dormitory. This door represented "preparing to graduate," and she practiced walking through it again and again. When she finally did graduate from Franklin Pierce College, she ceremoniously walked through yet another door in the college library roof. "Each door or gate," Grandin writes, "enabled me to move on to the next level. My life was a series of incremental steps."

Doors became an elemental symbol for Grandin, and she often associated them with wind. ("Wind has played an important part in many of the doors. On the roof the wind was blowing.") Similarly, although she had no real idea corresponding to "getting along with people," she was able to find an image that fit this strange notion. It was a bay window in the student cafeteria, consisting of three glass sliding doors enclosed by storm windows that Grandin had been asked to wash. In order to wash the windows, she had to crawl through the sliding door. Once inside, however, the door jammed and she was stuck inside; only by easing the door open very carefully was she able to exit from her temporary prison. "Opening the sliding door while trapped between two storm windows" became Grandin's guiding image when trying to pin down the rules of relationships. As a symbol, it enabled her to grasp a principle that would otherwise have been hopelessly abstract. In the same way, Grandin mentions a young autistic girl called Jessy Parks for whom the phrase "partly heard song" equaled "I don't know." At some time in her life, a partly heard song had become associated with not knowing.

It might be tempting to think of this autistic language as being comparable to the symbolic languages of those societies known as primitive peoples in anthropological terms. Concrete imagery forms the linchpin of an interpretation of the world that has symbolic depth; people who think in images might have an emotional syntax that is different from those who mash everything up with words. I had noticed already that Asperger writers are not comfortable with words at all. Their language seems stilted and wary, as if they are balancing on artificial legs over a floor of eggshells they are desperately trying not to break. It is almost writing without words at all, or using words as a kind of clumsy and frustrating telegraphic replacement for thought. For them, all language is a foreign one.

★

Many autistics had by now told me that the Internet was their savior, their only lifeline to the outside world. This is obviously true if one considers the mechanics of publishing. The Internet has spawned a literature of the Autistic Underground that would not otherwise have seen the light of day. Poems, novels, activist diatribes, political treatises all cram cyberspace in a way they could never cram a bookstore's shelves.

In some ways aut. lit. (as it might be called one day) is a genuine samizdat. Eager to track down Jonathan's literary colleague in Rotherham, David Miedzianik, I found a place where this gentleman posts his poems. Like Jonathan's stories, Miedzianik's poems are a record of the inner life of Asperger's shorn of the optimistic pieties of medicine and rehabilitation. I came to hoard many of these angry poems, with dour titles like "Get me some songs done so I can get some hugs," "I'm feeling sicker and sicker," and "I can't think of a title for this." The poet could also be seen posing for the camera in front of a red-brick semidetached English suburban house, a wounded-looking man of about forty with a red beard. Many of his poems are short haiku-like productions in the style of E. J. Thribb, that wholly invented poet laureate who so charmed the British in the 1970s with his four-line pastiches. David's, though, are fierce, relentless:

> The phone bill came on Friday and my phone bill was £126.
> That's what you get when you're on the Internet.

Another lament:

> Well it's New Year's Day 1998 today.
> There's no one to talk to and I can't think of anything to say.

Love is a constant worry:

> Why someone won't come to love, to take this pain away.
>> I haven't a clue.
> Well maybe I do have a clue why they don't want me.
>> It's to do with the autism.

The mood is one of unrelenting pessimism. In "I'm feeling sicker and sicker," David dreams constantly of getting a plane to Denver, while having to actually wake up in Rotherham.

> There's too much isolation for me in Rotherham,
>> and too many memories and pain.
> This is a town where no one will date me,
>> but most people know my name.
> They've such a downer on me that no matter what,
>> they won't have me for tea.
>
> I think I'll take myself to Denver the Mile High City.

There are three overriding concerns in Miedzianik's poems: girls, finding someone to have tea with, and autism.

> I'm all 6's and 7's with my sleeping these days.
> Mind you the obsessive compulsive disorder is a bit better now.
> I don't seem to be checking things as much these days.

At the launch of an Autistic Awareness music CD in Doncaster, the *poète maudit* at last manages to solve at least one of these dilemmas.

> Although at the album launch in Doncaster,
>> I did dance with an autistic lass.
> I paid £15 to get a taxi up to Doncaster that day.
> Although the woman from the Doncaster Autistic Society
>> did take me back home.
> That was one night I didn't feel so alone.

However, I was curious to go back to his poem about the St. Louis conference and see what he had done there. Had he enjoyed himself? I seemed to remember that he had met Temple Grandin and given her a book of his poems. It must have been an odd meeting:

When Temple Grandin finished talking, I gave her a copy
     of my new poetry book.
She signed the book, and gave it back to me.
So I gave her another copy of the book and she signed that too.
I told her she could have one of these signed copies.
So she took a signed copy of my new book, signed by herself.
I send her on all my books.
Although this time I met her.
So I gave her this one in person.

Well I was at The Autism Symposium in St. Louis again.
Some of the things they were talking about there
     were too hard for my brain.

Miedzianik has become something of a local celebrity in Rotherham and around Sheffield; the heavy metal band Solitary wrote a song
about him called "Twisted." In some ways, they explain his condition
more eloquently than he does himself, but there is undeniable pathos in
his lines. In one poem he more or less admits that words cannot depict
either memories or the inner autistic condition:

I can remember going into Derbyshire in my mum's car.
My mum used to take me for walks into Derbyshire a lot.
We'd go to cafes and we'd look at old churches.
Sometimes we'd go into Bakewell, and we'd walk about there too.
There was a river that ran through Bakewell.
I can't remember what this river was called.
One thing I can remember is this old café we used to visit in Bakewell.
There we'd eat Bakewell tarts and cream and jam scones and things.

I have trouble summing up all these memories with words.
Words don't sum it up, as much as being there,
     anyway it's all gone now so why care.

Years ago I always used to have trouble with writing and,
     putting things down.
Although this didn't stop me from getting around.
Years ago I used to ask, the young girls to meet me.
They never were much bothered and I used to go
     most places by myself.
Words were never much good at summing up how I felt then.
Instead I ought to have drawn a picture of my thoughts.
I never was much good at drawing then.
My mum and gran used to think I wasn't too smart.

I have trouble summing up all these memories with words.
Words don't sum it up as much as being there,
     anyway it's all gone now so why care.

There were other poets, too. I looked up David Spicer. His web page offered a soulful image of a sad-looking fellow with Dostoyevsky eyes. But his tone could have hardly been more different from Miedzianik's. His poems are always strictly rhymed. One of them was about a "diathermy machine." Spicer explains:

> My father was a podiatrist (foot doctor) for more than 50 years. An old piece of equipment he had was called a "diathermy machine"—it was the size of a small suitcase and had several adjustable coils and spark gaps. It produced RF (radio frequency) waves which were sent through the desired area of tissue to heat it up. It had laid, unused, in the basement for a long time. In my early teenage years, I asked my father if he could "fire it up." He did. Man, was it impressive . . . the spark gaps sizzled, the sharp tang of ozone filled the air, all TV channels were just obliterated . . . we didn't leave it on very long.

> *Diathermy*
> When first the ancient switch was thrown
> we sat in silence there, alone,
> but then, as air gaps were adjusted
> (years before folks got Ghostbusted)
> sparks did leap and dance and sear
> and scare us halfway to next year.
> I gotta say, it was dramatic
> drowning Channel 8 in static—
> was their program bad or good?
> Well, no one in *our* neighborhood
> could tell if it was even there
> 'cause Pop and me in basement lair
> were being Scientific Men
> consuming all the oxygen
> and making a tremendous fuss
> 'till Mom came and reminded us
> that maybe we had done enough,
> for one day, of nostalgic stuff . . .
> And so, the Grand Device was put
> away once more . . . in fact, for good . . .
> But still within my mind it lives,

and tissue-warming pleasure gives,
its Last Hurrah when Last Connected
gratefully now recollected.

I wondered what a diathermy machine could mean to a small Asperger boy, and why he would want to write poems about it. Even odder is Spicer's meditation on a coffee percolator.

Something over twenty years ago, I transferred from the Engineering to the Data Systems department at the phone company in Connecticut. Fresh out of training class, I was assigned to one of the work groups. As a new member, I was given the task of making coffee for the group. We had a commercial-sized machine, so it wasn't too much trouble. After a while, I started noticing a plastic pitcher of water next to the coffeemaker each morning when I came in. It seemed that I had a helper. I gratefully used this water for each day's first pot of coffee. Some time later—days, weeks, I don't remember—I overheard a conversation in the group. It seems that another new member had the job of keeping the group's plants watered, and was puzzled that the pitcher of plant food mix kept disappearing. . . . Not long after, I no longer had to make coffee . . . and so, again in commemoration:

Good to the Last . . .
I tried to make my cohorts green,
the healthiest green you ever seen.
I wonder if they would have minded
growing hair all leafed and vineded
losing fast their morning glower
bursting out with flower power.
(Slaving at machines of Babbage
looking just like haunted cabbage,
Punching cards of Hollerith
while smelling kinda Hyacinth.)
Lush, abundant women and men
getting all their nitrogen
While cautious, in this world of evils,
never to get et by weevils.
If, some day, their bloom should fade
and potbound skulls thin hairy glade,
not one would be a balding serf—
they'd just get rugs of Astro-Turf.

Spicer's poems had a quota of wit to them, and I was intrigued. Unlike Grandin's, Spicer's language was unvisual; it bounced along like that of a limerick or eighteenth-century satire. It was, in any case, quite different from the rather grinding prose he used to explain his psychiatric condition.

I discovered that Spicer lived in Asheville, North Carolina, a liberal arts city nestled in the the Great Smoky Mountains. His reasons for living in Asheville were not explained, but it seemed that he had retreated there, retiring from the stresses of the neurotypical world. Since people, especially poets, are intricately entwined with places, I was curious enough to travel to Asheville to find David Spicer and ask him myself. After all, I reasoned, the Autistic Underground must also have its special places, places where ordinary life was allowed to coexist with the poetry of diathermy machines.

★

Asheville is one of the strangest cities in America. Misty mountains surround it, seeming to cut it off from the rest of the South, enclosing it like a primeval wall, while within it a liberal student culture seems uneasily in charge. It was a humid Friday evening in May spattered with tropical rain. With nothing to do before meeting David at the local Barnes & Noble outlet, I wandered around a few hilly streets of brick charm, refreshed by a smell of wet forests. What, I wondered, had drawn Mr. Spicer to this enjoyably kooky town? The art deco buildings like Schotzy's, perhaps, or the dim medieval faces and griffins carved into a place called the Dr. Humor Building? I ventured around Pack Square, Asheville's corporate heart dominated by a tremendously hideous Merrill Lynch tower and a needle monument to the Confederate war governor Zebulon Baird Vance. From here the wooded hills looked stiflingly close. There was a desultory Linda McCartney exhibition at the Art Museum, but the sidewalks were almost empty, the fountain and geometric lamps abandoned to the rain. On this occasion, I decided to give the Colburn Gem and Mineral Museum a miss.

Instead I waited for David Spicer inside the Asheville Mall. We had made a rendezvous at the bookstore café, a franchise called

Republic of Tea. There was almost no one in the immense store, and I waited alone among the tea tins of Desert Sage and Zen Dream. Another customer at the counter earnestly explained the intricacies of the Constitution to a bewildered waitress, while gesticulating wildly with both hands. So this was Friday night in Asheville. I tried to read a magazine, while ignoring the Nissan thermoses and Bodum *cafetières* that no one ever bought, but I could not help perusing the shouting customer. Bright turquoise shorts, plaid shirt, white bobby socks and Timberland boots, a cup of Writer's Chai wobbling in one hand. "Political indoctrination!" he was bawling, swaying from side to side. "Government brainwashing!" Was this David Spicer the poet? "I'll take a Hugger-Mugger Brownie, too," he spat and slurped the Writer's Chai. Who but a writer would slurp Writer's Chai?

Dismayed, I prepared to slink away. But, as I was rising to do so, a pair of feet shuffled behind me, and a shy middle-aged man in a Special Olympics T-shirt came up as quietly as a ghost and held out a hand. So softly spoken I could hardly make him out, the true poet and the author of the ode to the diathermy machine announced himself.

"Glad you made it!"

Dave looked swiftly around the Republic of Tea and sized it up. It was clear at once that he came here often and knew his way around the Zen Dreams and Mango Ceylons. He sat and said at once, voice suitably lowered, "You know, people here are conditioned to accept maddening conditions. Look around you."

"At least it's quiet," I said.

"That it is. Why, that's why I come here."

He looked at me intensely, and I thought that something flickered deep inside his eyes, something not purely physical.

"It's what I call America Normal," he smiled. "When I'm up against America Normal, I feel kind of deliciously subversive. I call myself an autistic sociologist. Normalcy is highly overrated, you know. We have two eyes, so why can't we have two perspectives on social life?" I could see at once Dave was an intense fellow once he got going. "Those of us outside America Normal have a few comments to make on it. We could help the mainstream see itself for what it is, which it can't do most of the time. We're the canaries in the coal mine!"

"That would be very sociological of you."

"It sure would, yes. That's exactly the word. In fact, I'm getting more and more into sociology. I'm studying it at school. I'm doing Self and Society. Every autistic person is a sociologist. We have to be."

"What about America Normal?"

"Oh that word, normal. Why do neurotypicals always want to bring us down to their level? For example, what is a normal level of noise? This culture never thinks about noise. Is the noise level normal in New York?"

I thought of my sleepless nights in Brooklyn, broken up by the crash of *Mad Max* garbage trucks at 4 A.M.

"I see your point," I said. "But is it normal here?"

"Oh, Asheville's very quiet. My wife Dove and I moved here in 1981, because we knew it was a peace town. It's where *Mother Earth News* is published. That's why we came here. We're avid readers of *Mother Earth News*, see."

For an hour or so we talked about his life.

After doing his engineering degree, Dave had gone into computing while hankering after the beauties of the written word. After longtime problems with alcohol, he decided to both move south and explore the possibility that he might be autistic. He had noticed that his son Andrew, who was born in 1985, has ADD and Oppositional Defiance Disorder (talking back to one's parents). "And," said Dave disarmingly, "he has Obsessive Compulsive Disorder as well. So it naturally got me thinking about myself."

Writing was his salve.

"I have what's called a 'splinter skill.' That is, I can do one thing well. In my case it's verbal fluency. But I don't actually express anything with it. I just use it."

I had the impression that all the earlier parts of Dave's life had taken a backseat after the drama of his diagnosis around 1993; his life had subsequently been turned into a medical melodrama that has not been assuaged by the fact that Dove also has ADD. The family was a veritable melting pot of *Diagnostic Manual* conditions, though it had not prevented Dave from finding a nice quiet job in the county library system and completing his major in theater.

Does he have any job aspirations?

"I volunteered for Parking Enforcement. I wanted to get active about enforcing handicapped parking space. But I couldn't deal with people's indignation. I'm very adamant about rules, you know."

I thought to myself that Dave would indeed make a pretty alarming handicapped parking space enforcer.

"People said I was brusque. I was."

Dave is inordinately sensitive to noise, and I soon noticed that he was highly disturbed by the purring of the café's ice-cream freezer. He began to squirk and scowl at the fridge, eyeing it as if it were a lethal animal about to spring on him. I therefore suggested we meet the next day at his home, while he, in turn, suggested that I accompany him later in the afternoon to a meeting of the Autism Society of North Carolina, where he was giving a presentation. This sounded like an excellent plan, and we got up to go. As we did so, Dave scribbled his telephone number on a card. Then he looked at it askance and realized that it was not the right number.

"Why, look at that. I just wrote my phone number from 1979, from before I was married. Well, well." He tutted and rolled his eyes. "That's what you get from seventeen years of alcoholism."

★

Dove and Dave live in the Mosswood trailer park in Bunscombe County, just outside of Asheville. The unit is atop a large hill, with views all around, but Dave was inside listening to ZZ Top. On the coffee table lay a guide to *Babylon Five*.

"Generally," said Dave, "that's the only television show we watch. I have 110 episodes on tape. *Babylon Five* has taught me a lot about life. Really."

I flicked through the guide while Dave made coffee.

"Do you follow *Babylon Five*?" he asked.

The unit was decorated with flounced curtains, a frosted diner fan, and tired carpets. I shook my head. The Drakh reveals the keeper that will control Lundu in The Fall of the Centauri Prime. A new artifical eye for G'Kar. Byron's telepaths. Garibaldi has to confront his addiction to Buovari liquor. Guide to episode 5.

Dave now told me exactly how his extraordinary obsession with *Babylon 5* actually worked. What was important to him primarily was that the episodes of this sci-fi epic were spread out over five years. This made the slow evolution of human relations comprehensible to Dave for the first time in his life. It was a reminder, in his own words, that things take time. *Babylon 5* became a kind of televisual social manual, explaining with visual immediacy the nature of things such as "choice and its consequences, spiritual awakenings, sacrifice, standing up for one's beliefs, loyalty, having an active life, having compassion—that sort of thing."

It sounded like an onerous layer of meaning for a mere TV series. But Dave used *Babylon 5* as his personal bible, and missing an episode of it was like missing a Sunday sermon at church. When *Babylon 5* was temporarily taken off the local airwaves in Asheville, Dave panicked, then loaded his car with a VCR, a small TV, a UHF antenna, and a converter and drove ninety miles to Shelby in North Carolina. Now in range of the nearest station broadcasting the series, he would begin recording the missing episodes in order to keep the series' architecture intact in his mind. "I did the trip forty times. I have all one hundred and ten episodes."

There was a small commotion, and in came Dove. For some reason we began talking about food.

"We don't really eat food," said Dove, philosophically. "We eat fast food. Don't we, Dave?"

"Yeah, basically."

Dove told me the tale of her previous marriage to John. John was on Prozac for his ADD, while she was on Ritalin and Wellbutrin for hers. But John developed Tourette Syndrome while on Prozac and then acquired tics, throat clearing, and other generally mad behavior. Dove couldn't handle it, and they divorced in 1997.

"God," I gulped, "how awful."

"Oh John's OK," Dave put in. "In fact, he's coming with us to the autistic conference."

"Is that a good idea?" I ventured.

"You'll love John," said Dave warmly. "He's a character."

As Dave and I drove to pick up John, he waxed morosely philosophical about his various conditions.

"A recurrent theme of my life," he said, "is the intertwining of responsibility and opportunity. The process of dealing with it makes me grow. It's the emptiness and hollowness of American life, you see, all the distractions."

But at least that same hollow American life, I thought to myself, enables you to spend your entire life thinking and talking about yourself—a luxury of extreme rarity in the rest of the world.

"Rage," he went on, "needs not being met, acting out, frustration, quality of life. . . ."

I asked him about his Asperger son.

"Well, he has compulsions. His mother has clinical depression. But he's cool with it now."

Dave utters everything with a mechanical, eye-avoiding wryness that gives to everything around him a faint atmosphere of the Mad Hatter's Tea Party. This atmosphere only increased when John Englo got into the car. Dove's ex-husband was dressed in a hunter's cap, khaki battle fatigues, flip-flop sandals, and a green military shirt with leaf patterns. He looked like he was about to go on a fierce deer-killing spree with his bare hands, even if his manner was gently subdued. "I have a degree in biology," was the first thing he said. "Are you Asperger, too?" As we sped along the freeway in the direction of Black Mountain, our conference's venue, John confirmed that he had indeed been married to Dove and that he had indeed begun to suffer throat clearing, numerous tics, and the occasional sticking of his fingers down his throat as a result of a regime of antidepressants.

"It's all true," he said gaily. "I guess marriage just did me in. I used to sit up on roofs and hide under tables. I used to wedge myself behind the sofa and try to get a grip on the world. I needed to get deep pressure on my back, see. Then Dove told me to squirt the cat with a squirt gun when it misbehaved and that was just too fast for me, man. The cat and I were both traumatized. You can't make cats do rules. I'd be there rocking under the kitchen table with the squirt gun, paralyzed. Like I said, marriage did me in."

As he was talking, both he and Dave were rocking slightly as if in tandem. Dave explained that he now had four different syndromes—namely Tourette's, Asperger's, OCD, and full-blown depression. He was stressed right now because he had to do this panel, so he did some hand-curling to calm himself down. I looked down at his lap and saw that his hands were indeed contracted into curls. When his seizures got too bad, he went on, he had to lie down on the floor, and wait for sleep to relieve him: there were "ballistic contractions" and "dystonic contractions." Otherwise he was taking Paxil for his Tourette's.

"Does it work?" I asked.

"A psychiatrist would say so. I don't really know."

"What about the abdominal tics?" Dave said.

"Oh, they're cool."

"And the neck tics? What about the vocal tics?"

John laughed. "Yeah, I shout dada things, you know. Sometimes I throw myself right out of bed and start doing my dada tic."

"How does it go?"

"I just start shouting my mantra. Bah, manananana, bah, manananana. Something like that."

"Is that with the neck tic or without it?" asked Dave.

"Usually with. I have to be spoon-fed because I get my wrist tics."

"It's pretty bad, isn't it, John?"

"Yeah, it's pretty bad. Like I said, marriage—"

Dave nudged me. "But the worst thing about the wrist tics is that John can't play the banjo. He's a virtuoso banjo player. Aren't you, John?"

We arrived at the Black Mountain Lodge, a spacious seminar retreat much used by businessmen for rustic conferences. Among the stifling forests of pines, a few makeshift outbuildings on the far side of the road from the main house had been set aside for the autistics. Next to the meeting house, a large construction crew was setting up some caterpillars and drills for extension work on the complex. A large band of students—future counselors for autistic children—trooped into the hall in jeans cutaways and sweatshirts, tanned and childishly robust as American undergraduates often are. They made for a striking contrast with the panel of autistic adults.

Apart from Dave and John, there was a nervously aggressive girl of twenty-three called Maria who demanded total silence from everyone, because she was exquisitely sensitive to noise. Next to her sat a relaxed middle-aged Asperger man with a pale smile named George. George, called upon to describe his past, immediately announced to everyone that he had graduated from college in "refrigeration and air-conditioning studies."

Maria stood up and more or less shouted, "If you'd be so kind as to turn off all your mobile telephones, please. And no coughing, please. And no applause. Any sudden sound is excruciating for me. You have to understand . . . it's physically painful, it really does upset me terribly . . . so no phones, please!"

Suitably intimidated, the students fell into complete silence. John began talking about his love of banjos.

"I almost stopped college," he began, "because it got in the way of my banjo playing. There are forty-seven different tunings for the banjo. . . ."

The mastery of detail pertaining to narrow subjects brings out a characteristic Asperger intelligence, according to which a magnified foreground obliterates the wider view. In a sense, this is why an Asperger writer could not write a good novel, or even read one. The wider panorama is too slowly revealed. I began to realize that John's inner universe was calibrated not by his recent marital disasters or by his propensity to ram his fingers down his throat while sitting on lonely rooftops, but by the forty-seven different possible tunings of the banjo, all of which he had memorized aurally by listening to old recordings of masters like Mike Seeger and Paul Brown. There are subtle vibrations that inform the ear which tuning is being used. John learned these vibrations by turning his entire front room into a sound chamber, piling everything in the house behind the sofa to deflect the sound, and then plucking single notes on his banjo until he had learned the exact vibration. This exercise, endlessly repeated, had turned his brain into an inexhaustible library of banjo tunings.

"And," he added forlornly, "totally useless."

The audience applauded silently using the deaf applause hand-sign, wriggling their fingers above their heads, before the other speakers offered their autibiographies.

Maria: "I come from a family of wine connoisseurs. My father was a violently abusive alcoholic."

George: "I collect coins. I own forty computers. I drive a Volvo."

John: "I carry earplugs wherever I go. I walk with my eyes closed. I wear the same outfit all the time. I only eat Snickers bars. It's remarkable what you can do and not die."

Maria: "I prefer coffee to food. Touch can be like an electric shock for me."

George: "I've learned to lie. But I can't lie and look away at the same time."

Dave: "Autism is my life. It's what I do. It's interesting stuff. Chance to grow . . . quality of life . . . win-win situation . . . my dad was an alcoholic. . . ."

It was like one of those unconsciously competitive conversations that children have. Boy A: "I broke my arm"; Boy B: "Yeah, well, I was in a traffic accident and split open my skull"; Boy C: "My dad split open his skull and his rib cage!" and so on.

Everyone chimed in at the same time to claim that their families had all been alcoholics. The students nodded, frantically scribbling, then raised their wriggling fingers in deaf-mute applause. A hidden mobile phone, however, suddenly went off, and Maria jumped to her feet again with her hands clasped to her ears.

"No! No! I said no mobile phones!"

George raised his hand. "The crackling of potato chips is what annoys me most. I eat oatmeal with molasses because it's soft. Red and yellow put together really get to me. . . ."

By now I was beginning to feel subtly restless. Something about the hall had become claustrophobic to me, and I began to perspire. I squirmed in my seat, not knowing exactly what it was that bothered me so much. Perhaps it was the total silence enforced by Maria, which reminded me uncomfortably of school. However, this same silence now came brusquely to an end. As Maria was talking in a hypersensitive whisper about her alcoholic father, the construction crew next door suddenly burst into life. A devastating cacophony burst over the proceedings as four pneumatic drills and the bulldozer went into simultaneous action, and the windows began to shake as if an earthquake had erupted beneath us. Stunned, I looked at Maria. But she had continued talking, apparently unconcerned; and none of the other panelists had reacted in any way whatsoever.

The uproar was so brutal that within ten seconds I felt that I was losing my mind. I crouched in my seat with my hands over my ears while Maria continued what was now a mime's speech of which I could not understand a word. The windows shook; the ground trembled. Now it sounded like an artillery barrage. The speakers gazed gently into space, unfazed.

It was at this moment that I realized how estranged from them I really was, when all was said and done. If the Asperger person feels like an anthropologist on Mars, then I now felt like a Martian who has accidentally walked into a convention of anthropologists. My sweating became more violent, because in fact I suffer from an inordinate sensitivity to noise, and the jackhammers causing the frame of the hall to vibrate also seemed to be pounding symmetrical holes into my cerebellum. While Maria carried on talking about her father with complete poise and equanimity, I rose unsteadily and looked round for a discreet way of exiting the scene. I at once felt cumbersomely obtrusive, ill at ease, as if my legs were all tangled incongruously with my arms. Interestingly, I would say that I felt as I often did when I was a child, and it is not a sensation that comes to me with any frequency. As I tried to slink unnoticed out of that meeting of the Autistic Society of North Carolina, I could not help thinking of my sly escapes from the birthday parties of my childhood. Flight in solitude, a nameless panic. Only afterward in the parking lot did I meet up again with Dave to say goodbye, and I raised the question of the atrocious noise coming from the construction site.

"There's nowhere quiet in this country," I complained.

But Dave shrugged, and even looked a little nonplussed.

"Oh I've heard worse," he said. "Did you hear that goddamned mobile phone that kept going off?"

"You have to be joking."

"No. Bach. I do hate it when phones have Bach tunes."

"I didn't notice," I confessed.

In the humid air of the forest clearing the shattering artillery barrage continued. Dave smiled sweetly and shrugged; he looked far less lost than I felt, if the truth be told.

"Well," he said with a sage finality. "We're not all sensitive to the same noises are we?"

"I guess we aren't."

"That's diversity for you."

We shook hands and I asked him if he was writing any new poems.

"Not right now," he said fatalistically and sighed. "Right now I'd like to try once more for Parking Enforcement."

CHAPTER 7

THE POETICS
OF MEDICINE

That was the point that Razak—who was awed by the djinn—
struggled to make with his English. He had seen two or three other
people possessed by djinns, he said. But then he said that he was sure
that in other countries, other civilizations, people would believe in
other things, mental illnesses would take other forms, and there would
be other cures.

—V. S. Naipaul, *Among the Believers*

Lundu lies on the border of Indonesia and Malaysian Sarawak. It's a
hot Chinese market town lost in the jungle, nourished by the produce
of the Iban and Dayak farms that run along the northern coast of
Sarawak. A few miles out of town, I sat inside the bare house of a
Dayak grandmother named Dibuk ak Suut, a Catholic matriarch of
Kampung Sebiris, whose mummy-like body was wrapped in batik. Her
front room overflowed with the complicated paraphernalia of Malaysian homes: government certificates, ceremonial swords, teapots, and
painted wooden fruit. She sat on a mat watching a Chinese wrestling
match on a huge Aiwa TV, while her husband Sujang served us tepid
hot chocolate in glass cups. They were farmers in the humble business
of coconuts and pineapples; machetes lay neatly stacked against the
back door. The jungle came right up to the windows, its insects shrieking. Sujang winked at me and nodded toward his wife, a tiny woman of
wrinkles and twitches in a leaf-green sarong.

"Observe, please," he whispered and got up.

My Dayak interpreter from the capital Kuching looked a little alarmed and crossed his legs. "Observe, please," David duly said in English.

Sujang crept up behind his wife. Merriment all round. I assumed a lighthearted family prank was about to commence, but would it be malicious or harmless fun? Sujang clapped his hands, and Dibuk, without warning, shuddered, her face twitching in time with her twitching fingers. Her eyes went glassy and she rose to her feet, her hips slinking from side to side like an amateur stripper's. Sujang could hardly stop himself from exploding with laughter. He went up on one leg. As if hypnotized, Dibuk mimicked him. Then Sujang made a grotesque exaggerated wink with one eye, and tiny Dibuk did the same. She seemed to have lost all control of her mind or limbs and her childishly asymmetrical face went into a trance.

"Grasshopper!" she shouted. "Grasshopper!"

David and Sujang began snorting with hilarity.

Dibuk began hopping up and down on one leg, with her face contorted in the Popeye grimace. She now noticed the black family cat staring up at her in terrified astonishment. She shrieked again:

"Cat black prick! Prick, prick, cat, black! Cock!"

Having translated this for me, my trusty interpreter looked distinctly alarmed, and I had to calm him down. I was there to meet Dibuk because she suffered from an obscure mental disease known in Malaysia as *latah* and in the West as "hyperstartle syndrome." A latah, as someone afflicted with this rare illness is known, is usually a middle-aged or elderly women easily startled by sudden noises or surprises, whether they be falling coconuts, backfiring cars, stealthy cats, or handclaps. Whereas you and I will startle only for a second or two (while swearing, going into absurd convulsions, or shouting), a latah will remain in this trance-like state for half an hour. During her trance, she will imitate everything around her, obey all commands however indecent, and utter obscenities that seem to come from the depths of her unconscious. Afterward, when the startle trance has been broken, she will have little or no memory of her absurd behavior. No one knows what latah is or why Malaysian women are so prone to it; the only comparable syndrome known to Western medicine is a condition once common among Canadian loggers, who became hyperstartled by the crashing of falling trees. It is officially known as "Jumping Frenchmen of Maine Syndrome."

Sujang leaned over to me and poked my shoulder.

"It's cats that startle her the most. She can't deal with cats."

"Does she imitate what's on TV?" I asked.

"Sure. The wrestling shows are a big problem. We have to hold her down. Not good, not good. But funny."

Later, Dibuk "woke up" and blinked her eyes.

"My goodness," she piped, "did I say anything obscene?"

"You said cat black prick, Dibuk."

Dibuk buried her face in her hands and laughed. "No, no!"

"Yes, yes you did. The cat's prick, you said."

"Well," she said serenely, "I've been latah for thirty-five years now."

"And how did you become latah?" I asked.

"I was poked repeatedly," she said gravely.

Gilles de Tourette, the late-nineteenth-century discoverer of the verbal tic that now bears his name, often compared it to latah, which was well-known to European medicine, even though its causes remained baffling. What is clear, however, is that in Malaysia interaction with latahs has become a complex form of social play. Rather than being despised or cast out, latahs are accepted, even actually celebrated, for their oddity.

Other Malay syndromes were also much studied by colonial doctors, especially *amok*, the murderous fits that came over brooding, sexually jilted men in villages, and the infamous *koro*, a delusion suffered by men who are convinced that their penises are shrinking inside their bodies. I had come to Malaysia, in fact, to see if these mysterious and exotic mental illnesses still existed. Ever since reading Stefan Zweig's terrifying short story "Amok," I had wondered about these bizarre compulsions: were they real afflictions, or were they picturesque inventions of a condescending colonial psychiatry that had actually reached a high degree of investigative sophistication in British Malaysia? The British, in fact, had taken a keen interest in indigenous mental disorders, perhaps because they lay at the root of many of the domestic crimes, which they were obliged to investigate and judge. Legal judgments against men who had run amok, for example, had to respect amok's cultural context or risk seeming unfair.

The 1994 *Diagnostic Manual* not only included Asperger's for the first time—it included latah for the first time as well. The latter, it explains, belongs to a curious group of illnesses known as "culture-bound syndromes." These are mental disorders that appear to be created by purely cultural rather than pathological forces. Culture-bound syndromes may be rare and exotic but they're also controversial, for they raise intriguing and profound questions about the very nature of mental illness.

As we have seen, Western psychiatry has become increasingly dominated by a biological model of mental disorders. This has resulted in modern psychiatry becoming more and more rooted in the hard sciences. Nature, says modern psychiatry, is the hidden motor behind mental disease, and medicine is required to treat it. Moreover, the biological model also implies that mental illnesses are universal afflictions and that, in the same way that people around the world get cancer or diabetes, others get delusional or depressed.

Yet recently many psychiatrists have come to question this model. Why, they ask, do so many American women get *anorexia nervosa*, while Indonesian women almost never do? The fact that certain syndromes blossom in some parts of the world but not in others suggests that the ways in which a mental illness manifests itself can be shaped to a remarkable degree by a person's environment. Indeed, can culture actually *spawn* mental illness, afflicting people who, in other environments, would be perfectly normal? Many who study culture-bound syndromes increasingly claim that it can, and have set out to describe the subtle threads that sometimes connect a culture to its peculiar psychiatric derangements. Meanwhile, a kind of encyclopedia of world disorders is emerging from this research.

In consequence, bizarre disorders have come to the surface. In West Africa, young males suffer from a disability popularly known as "brain fag," characterized by a chronic inability to concentrate in university libraries, combined with blurred vision and burning sensations. In India, men suffering from a disease called *dhat* become hypochondriacally anxious with their semen discharges. In Latin America, people fall victim to a sinister mania called *susto*, in which people who have been traumatized believe that their souls have been separated from their bodies. In a Nigerian outbreak of koro, widely reported in the 1990s, hundreds of men became convinced that "penis thieves" were making off with their private parts. The theft could occur merely through a casual handshake; a wave of penis-theft hysteria swept Nigerian towns, with penile miscreants being locked up by the authorities pending a proper investigation. Such were the passions roused that several were lynched by their "victims" upon their release.

All this makes many Western doctors very uneasy. In a 1999 article published in *Culture, Medicine, and Psychiatry*, Arthur Kleinman, a medical anthropologist at the Harvard Medical School, argues that "massive global change" has created unprecedented conditions for the spread of ever more baffling mental illnesses. These conditions range from rapid urbanization and "transnational advertising styles" to changes in diet, fashion, and the media. As traditional cultures crumble in both the developed and the undeveloped world, the forces that hitherto held individuals together begin to dissolve, with unforeseeable consequences. "Viewed in mental and social health perspectives," he writes, "all is not well with global social change. And yet, we really do not understand fully what these changes actually mean with respect to everyday social experience at the level of communities, social networks, families and individuals." He continues:

> The celebrated historian Eric Hobsbawm, in a particularly bleak forecast of what lies ahead in the new millennium, observes that we are entering a period of so many changes with such powerful effect that the existential element in human experience itself is transmuting so that neither long-standing societal values nor established patterns of sense and sensibility are like to be the same, or act as a stabilizer of future forms of social life.

One baffling characteristic of culture-bound syndromes is that they can often take the form of social epidemics. In the late 1960s, for example, a wave of koro swept over the Malaysian peninsula, more or less as a typhus epidemic would have done. Doctors in Singapore were confounded. Why, they asked, were Chinese men complaining of koro at the very time when race riots between themselves and Muslim Malays were breaking out? Was it a kind of mass racial anxiety attack? At the very least, such a purely mental epidemic would suggest that some syndromes can be infectious in a purely social or imitative way rather than being anchored and isolated in the psyche of a sick individual.

This does not mean, of course, that modern koro or latah sufferers are consciously faking their affliction, any more than American girls, desperate to resemble their skinny peers, are faking an obsession with food.

As Kleinman noted in *Rethinking Psychiatry*, his groundbreaking 1988 book, "mental illnesses are real; but like other forms of the real world, they are the outcome of the creation of experience by physical stuff interacting with symbolic meanings." In Singapore I went to visit Dr. Ng Beng at the Woodbridge Institute of Mental Health, who explained to me that Eastern and Western conceptions of mental health were disturbingly different. Syndromes were not simplistically universal.

"Modern psychiatry," he said, "is essentially a Western import. In the East, people don't tend to distinguish between mind and body in the same way. My patients rarely talk about their moods *per se*—they always talk about physical symptoms." In other words, the way that a Chinese patient will conceptualize even a biological condition like schizophrenia is utterly different from the way a Westerner would "feel" it.

Latahs fascinated me because they suggested that mental illnesses might start from a neurological basis but then fan out, as it were, into a kaleidoscope of differing conditions, which were themselves the unique creations of different cultures. In 1968, the anthropologist Hildred Geertz argued in a paper called "Latah in Java: A Theoretical Paradox" that the sometimes raucous behavior of those with severe latah could only be understood in the context of the severe courtly restraint that is the norm in Malaysia and Indonesia. In these cultures, emotions—especially sexual ones—had to be strictly disciplined, and latah consequently arose as a kind of emotional outlet, a safety valve. Geertz also reported that latah was hardly an isolated case, for latah-like conditions seem to appear in other cultures as well. In rural parts of the Philippines, for example, a nearly identical condition known as *mali-mali* is widespread. In Siberia, there is a hyperstartle complex known as *menkeiti*, while in Thailand it is known as *baah-ji*, and in Japan, as *imu*. And then there are the French Canadian loggers, with the hint of a *Monty Python* routine in their syndrome. It is difficult to say whether these various hyperstartle conditions are like latah or, indeed, *are* latah. And, one needs to ask, would it be helpful to define a spectrum of latah disorders?

These variants not only make latah very tricky to fathom; they also raise the question of whether other syndromes might likewise arise out of some kind of social mimicry. The implications for medicine's understanding of mental illness in general might be far-reaching. Any given disorder or syndrome, far from being seen as a static, unchanging universal phenomenon, could be seen instead as a dynamic confusion of biology, infectious mimicry, social reaction, and even play. There would be no global standard for psychiatric normality, for although abnormality might stand out everywhere, Chinese normality would not be quite the same as American or British normality. Latahs, after all, are rarely considered "abnormal" in Malaysia even though their behavior would get them arrested in thirty seconds on Fifth Avenue. But even inside Malaysian culture, latah seems to morph as it seeps from one part of society to another.

Over a couple of weeks during my stay in Malaysia, I hung out with the latah gays and transvestites who throng the waterfront markets of Kuching. All of them are latah, or claim to be. Latah for Malaysian gays is a way of breaking down social barriers in a hostile Muslim environment, enabling them, for example, to pick up straight men without shame. A debilitating mental illness in elderly women had become a form of rebellious spiritual theater for young gay men: exceedingly strange. Kuching's most famous gay latah is an antique dealer by the name of Nelson Tan, who runs an upscale boutique on the waterfront—a formidable den of Dayak spears and Malay teapots. Nelson sat me down in his office and flexed his blue fingernails; a lithely boyish body gleamed through a cheesecloth shirt held together with pearl buttons. His manner was studiously queeny, almost caricaturally so, with many a flourish of the blue nails, rolling of eyes, and Chinese-accented sentences like "Oh latah, darling, so lovely, me latah, he latah, we always laughing, I *love* it darling."

I asked him if he *liked* being latah.

"Oh darling, *course*. No be latah so boring!"

"Do straight men laugh at you?"

"We all laugh at latah. But it make relax."

Later, he showed me the transvestites out at the airport waiting for male dates, discreetly removed from the public eye.

"They all latah, too. If you talk them, latah straight way. So lovely!"

A few hours of having tea with Nelson made it clear that in this radically different milieu—male, gay, marginal—latah was no longer a middle-aged female hysteria provoked by stifling conventions: it was the vehicle of a mild male hysteria provoked by the need to connect with other men. However, as I drove around town with Peter Kedit, a prominent Dayak anthropologist and former director of the Sarawak Museum, he explained to me that through latah both these marginal groups staged symbolic rebellions against the social order. Women masculinized themselves; gay men exaggerated their femaleness. "It's all about penis," he concluded gaily. "Penis, penis, yes, it's all sexual theater."

Was Nelson an eccentric? I couldn't decide. Are latahs eccentrics, or are they just latahs? Nelson's eccentricities were simultaneously natural and exaggerated, but one couldn't tell where the naturalness ended and the exaggeration began. Inside the hermetic individual, perhaps, it didn't matter. Nelson was a phenomenon, and people just left him alone. His deviation from Muslim normality was seen as amusingly perverse but never as pathological.

Gay eccentrics, "queens," like Asperger's people, often ferment in our imaginations in this way. We cannot tell where their supposed biology and their flowery manners begin or end. When I was growing up, the great Official Eccentric of English culture was the gay iconoclast Quentin Crisp. We devoured his filmed biography *The Naked Civil Servant* on TV (in which he was brilliantly played by John Hurt) and relished his hilarious sending up of the moribund English middle classes. With his dyed red hair, male prostitute antics, and languidly aristocratic verbal repartee, Crisp was my first TV hero, though I was not gay myself. He was an ethereal and wan Oscar Wilde for the 1940s.

Crisp's gentle and witty one-man war against heterosexual disdain was far more entertaining than today's dreary sexual activism, especially in America. It was clear that he hated collectivist "identities" as much as he hated church. But Crisp came to America "because it was more like the movies than the movies." And as an old man he became not so much the Official Eccentric (America does not want one) as a kind of Puckish lost soul inhabiting the New York demimonde, where his dark wit went mostly misunderstood. One night he showed up at a birthday party I was hosting on the Bowery, blown in from the street by a vague third-hand invitation, and sat himself regally in his flowing cravat on the end of a sofa. The Official Eccentric was now a frail old man with silvery hair, a strangely moving sight. His far-off manner has often since made me think of young Asperger's boys, and I had the feeling that this man had spent his whole life in almost total solitude.

"Would you like a drink, Mr. Crisp?" I said.

And I realized at once that no one ever called him *Mr. Crisp*—that thousands of bantering strangers would always call him Quentin, even though they had no idea who he was. He was simply Quentin The Queen, an amusing court jester identifiable purely because of his outrageous clothes.

So he laughed. "If you can *afford* one," he said.

Years later I would often see him sitting forlorn and alone in the window section of a dull all-night diner on Second Avenue, alone in front of a plate of fries with his velvet hat skewed rakishly to one side. I never quite had the courage to go in and say hello, and that Edward Hopperish vision filled me with a claustrophobic anxiety. What is an eccentric, after all? A man who despises normality and who lives by his own rules? And how does he live when his amusing side is no longer so amusing to the neurotypicals? Perhaps he must become that old man sitting alone in a Second Avenue diner, a shadow of a shadow of a shadow.

In any case, although I had not intended to draw comparisons between things like latah and mental syndromes in the West (even mere eccentricity), I could not help wondering if conditions like Asperger Syndrome were not also similarly complex. I had already noticed that the diagnostic criteria for Asperger's were so complicated and so contradictory and so blurred at the borders as to sometimes stretch credulity. There seemed to be endless lists of ifs and buts. Was it possible, then, that around a core biological illness a large superstructure of behaviors and moods had been created by the society itself? This didn't mean that Asperger's was imaginary, any more than is latah. Both are all too devastatingly real. But the purely biological explanation was too simple.

Many hold out hope of a cure for autism in the conventional sense, but so far such a panacea has proved stubbornly elusive. We might ask whether a personality disorder should be cured at all—for what would it mean to *cure* a personality? Do we even really know what a personality is in the first place, and by what impertinence do we affect to lay down its laws? A personality is not created the way a wedding cake is, by design, unless we are talking about the engineering of human souls in totalitarian paradises. In fact, no similes or metaphors exist for the personality of the latah or for that of the Asperger person. The latter are the unconscious creations of a lifetime. They cannot be undone. Nor should they be undone.

But whereas the latah is almost always accepted for what she is, becoming with time the center of many festivities and much ritual horseplay, the Western Asperger's person is usually totally isolated. And these two very different fates will always have untold consequences. Medicine will not often look beyond its own means when considering cures for personality disorders; it's understandably reluctant to venture beyond medicine itself, for beyond it lies much quackery. But the Asperger's life cannot be treated as a physical cancer. Much of its eventual course will be determined by a simple but terrifying fact: will isolation endure or be overcome? For I had already noticed by now that the agony of the Asperger's isolate is his lack of work and his lack of play—because one naturally produces the other. If Asperger's is partially a culture-bound syndrome, which it may well be, then the curing of it must also entail a great effort of the culture itself.

★

A few months after my trip to Indonesia I was in Montreal, driving alone through the Côte Saint-Luc on my way to visit an Asperger's woman named Marla Conn who had been writing to me for some time. I had finally decided to accept her invitation to visit Montreal, for Marla never left Montreal, she told me in her terse and sometimes almost angry e-mails, because the mere idea of leaving Montreal was enough to give her an anxiety attack. Marla, a fierce recluse, was as attached to her native city as Dibuk was to her native Kampong, which she also never left. Thus I found myself driving one evening through the Côte Saint-Luc, one of Montreal's vaster neighborhoods, toward a destination that was at least as exotic to me as Kampung Tebaro, if not more so. Marla lived in the Jewish part of town, a place called Cavendish; she had written to me that I would be able to get my bearings from the pale blue Israeli flags flying from the top of the Cavendish Mall. "You can't miss them. They'll make you laugh. You'll know you're in the right place."

As I spotted these same flags, I wondered to myself what Dibuk and Diyuk would think of these sidewalks, the upright signs crying *Marche vers Jerusalem*! Or what they would think of the front-page article in that day's *Journal de Montréal*, which I was reading at the traffic lights, and which told of a mother's murder of her fourteen-year-old autistic girl. Would they be able to make sense of a mother killing her daughter, because the latter was "eccentric"? More to the point, would the merry latahs be able to stay sane within the confines of this disarticulated urban architecture—among these long, looping boulevards that seemed to go on forever and that would probably strike them as febrile? I wondered if they had any idea of how lonely the northern winter can be, of what the Idea of North could mean at its very worst. I wondered, too, if they could imagine the inside of a modern Western apartment or even the interior of an apartment-building vestibule. If researchers of culture-bound syndromes think that every landscape has its illness, what would be the relation between Côte Saint-Luc, the Israeli flags on the Cavendish Mall, and the forty-something Marla trapped for a lifetime in her high-rise apartment on Norwalk Drive? Would her affliction be universal, subtly Montrealean, or simply North American?

Marla's neighborhood had a distinct feel to it, wide open and suburban. As befits a francophone city, the street names are cerebral: Sir Walter Scott, Honoré de Balzac, Einstein. Norwalk was a silent cul-de-sac and the lobby of her building a hushed arrangement of Persian vases, carpets, and sheet marble. The names on the buzzers all seemed to be Jewish. After a few minutes a pale, quite attractive woman came to the glass doors. I was, as usual, expecting the worst, because Marla had told me that she likes to hide inside cupboards and do math and that she also has Tourette's and "tends to bark" when agitated. But she was not barking now, and she didn't look as if she was about to dart into a cupboard. All she said was, "It's bright out here." We went down a long corridor, Marla telling me softly that strangers normally made her nervous, if not downright terrified.

"You're not a weirdo, are you?" she asked.

"Not at all," I said gaily.

"You don't look like a weirdo. But then again, most weirdoes don't look like weirdoes."

"I'm not a weirdo."

"Well, I guess you're not."

The apartment was modern and as dark as Jonathan Mitchell's had been. I noticed at once gray carpets, strange abstract pictures pinned to the walls, lowered shades, piles of *Woman's World*, and crossword puzzles on the sofa. A tense neatness ordered it all. Marla thought it opportune to inform me that this was so, because she no longer vandalized her apartments, especially not this one, which had been donated to her by her parents. In the gloom, Marla showed me her files room, which was filled to bursting with her private library of documents relating to (of all things) radiation, especially the literature of trace amounts in food. I could tell at once, in fact, that she was inordinately concerned with, and obsessed by, food, even though her kitchen did not seem to be a place where food was actually welcome. No knives, blenders, or spice jars, only boxes of Bazooka gum, containers of carob powder, *tartes au pouding* and bottles of cough syrup. The cough syrups were Marla's preferred breakfast. As soon as we got talking, she felt the need to explain the difficulties she was having with local supermarkets, which were, curiously enough, in some way connected to her problems with climate change. "How so?" I asked.

"Well," Marla said with surprising bitterness, "I must say that I hate the '90s, I hate the Millennium, and I hate change. More change is occurring these days—the weather is definitely becoming more erratic. Moreover, the food specials are disappearing."

"The food specials?"

Marla suddenly looked fiercely exasperated and flustered. She sat tensely on the edge of her sofa and clenched her fingers together until they were white.

"You used to be able to get food specials. Now you can't get food specials anywhere—they're disappearing. There are no gum specials—amazing! Sometimes I go into the supermarket and I attack bread when it's not on special. I feel that it's against my principles when it's not on special. So I'll attack it and break it open with my fingernails. I've been known to attack a fair number of loaves of bread in one go."

"You feel your rules have been violated?"

"Exactly. I have my principles, you see. I have my rules. Really, I feel a need for revenge against today's economy. One simply can't eat cheap anymore because of the demise of the specials. At the supermarket, it drives me mad, so I rebel. I'll hide frozen yogurt behind piles of cans. I call them French pigs and now I'm banned from even going in there. But I mean, why can't I get low-fat cottage cheese on special when I need it?"

"So you feel angry because it all seems like an enforced change?"

"Yes. Like I said, I hate change. I hate the weather here, and I hate the food. Sometimes I take to the cough syrup, get into bed, and throw the day away. I get very depressed. I'm terrified of storms. I hate Judaism too, by the way. The rituals get in the way of my rules."

Marla's complaints came thick and fast, and I began to wonder how ordered they actually were in her mind. What interested me was that she was radically removed from the normal regimens of social integration. She was a therapeutic outcast who had refused all the usual behavioral methods of treatment. Consequently, her medicalization was purely oriented around drugs. "I don't have the drive for self-improvement," she said dourly, "which other Asperger's people have. All this social training crap makes me sick."

Looking around her claustrophobic lair, I had the impression that Marla almost never went out or met with others; her life was a ceaseless war with others, especially the French-speaking citizens of Canada, who seemed to represent for her a unique form of evil—a conspiracy aimed at suppressing food specials and cheap gum. She admitted plainly that she didn't have human feelings, didn't miss her dead sister, didn't want sex or men, didn't want to be normal—but when frost damaged the vegetables in her garden, she'd cry for hours. She went to a therapist for a while, but he could not, she said, quite relate to her shopping needs. "I talked about my inner life and my shopping, but I feel he didn't quite get it. So I stopped." Marla's therapist, she went on, didn't understand that the world was changing, that supermarket prices were in flux, that the climate was unstable, and that the French-speaking citizens of Canada wanted to break away and form their own country. "Swine!" she burst out. And then she resumed the long list of things she could not abide, which caused her violent revulsion: doctors, specialists, psychologists, body contact, the thought of sex, Jewish holidays, the 1979 Soviet invasion of Afghanistan, the propaganda of handicapped martyrdom, humanities subjects, friends, isotopes bearing the number 5, and Montreal. But she loved numbers.

"Oh, yes. I love numbers. Numbers for me are a replacement for people. I hate people."

"Do you hate people more than you hate French Canadians?"

"About equal," she said tersely.

"What about jokes?"

"I hate jokes."

I asked her about her B.S. in math. Had she ever been able to get herself a math-related job?

"Never. My only real job was working in a hospital. I was put into housekeeping, but I kept chewing up my uniforms. And I hate bosses. I used to hide in cupboards. Then I got a job operating the lifting machine. Well, I soon became totally obsessed by the lifting machine. The lifting machine was my life. I would get up at five in the morning to tell the hospital staff about this machine, which they knew all about anyway. It got to the point where I began to hate any patient who didn't need lifting. I'd manipulate the staff into trading shifts with me so that I was always using the lifting machine. If I didn't get my way with the lifting machine, I'd start barking."

By now I was getting curious as to what Marla actually *liked*. When put to her, the question seemed to jolt her out of her psychological rhythm. We were sitting there in the dark with one lamp on, Marla rigid in her T-shirt, talking like a machine that occasionally shudders from the lack of an inner lubrication, and I had the sense that all the time she was talking, a kind of parallel voice was running inside her head. This was odd, because in response to my question Marla immediately said that she liked using her "private language," a lexicon of made-up words that she had been using since she was a small child. This private language offered her a large invented vocabulary that no one else understood and that, when repeated to herself, comforted her for her many cruel alienations. She never spoke them, she said; their power consisted precisely in the fact that they were never uttered, not even in solitude. They were "mind words."

"And," I asked, "is there anything else you like?"

Marla thought hard. She was deadly serious:

"Well, there's the cough syrup. And I like watching kids cry when they're getting syringes in the arm. I also like climbing up Mount Royal. And I like ice-skating. I'm a good ice-skater."

"Do you ever open the windows"

"The windows? What, for the view? I hate the bloody view. The bloody view is what I hate most about the windows."

"Do you have a private word for window?"

"Of course I do. But I won't say it. If I say it, it won't be a private word anymore, will it?"

I looked at the walls again, where there hung a few needlepoint pictures. When I asked about them, Marla told me that her mother had given them to her. Her mother? "Oh", Marla said airily, "she was burned to death in an accidental fire a while back." There was little remorse in her voice, despite the fact that her mother had provided her with all the furniture in her apartment. Marla was both dependent on her family's money and simultaneously unable to feign an emotional interest in them. For her, they were like the moons of Saturn. But this was not rejection or hatred, it was more like a failure of neurological wiring she accepted with passive helplessness. She had been more connected to her hospital lifting machine than to her mother or her sister. On the other hand, there was her poetry. But Marla wouldn't let me see her poetry, even though there were stacks of it piled on the kitchen table. Instead, she gave me something that she considered even more personal, namely photocopies of her medical records from when she was a toddler. "Read this," she said firmly. "I find them very interesting, anyway."

I wasn't sure quite what to say, but I read the Screening Observations for Marla Kahn, aged three, with considerable interest, for they were sure to demonstrate the way that clinical diagnosis had changed over the last forty years. The first report was dated September 23, 1959, and the name Kahn had been loosely crossed out by Marla herself and replaced by the more anglicized Conn.

The psychologist, Gloria Cherney, noted that Marla had not spoken before the age of two years and four months, and that she did not reply to questions, even when they were addressed directly to her. She separated easily from her mother and, the doctor observed, "easily took to the suggestion of making birthday cakes with Plasticine and sticks." The doctor suggested putting three candles in the cakes to reflect her age, but little Marla insisted on putting more and more, though Cherney did not suggest that the concept of number did not exist for her. When the doctor brought her something new to play with—water, for example—Marla smiled with pleasure, "but the smile was inward and not shared with us. She sometimes moved her lips without making any sound, as if talking to herself." She refused to play with a ball, evidence according to the doctor of "lack of interest in interpersonal relationships." And Cherney added that "her body movements are generally stiff and tense." She actively moved away from any physical contact. Her only signs of affections were directed toward ribbons, towels, and blankets.

It was clear to this doctor that Marla was autistic, even if she did have "a fairly good intellectual potential." The conclusion is spry: "She ignores people, has echolalic speech, uses one's hand as a tool and is sensitive to outside sounds. We feel she would be a good candidate for the Day Treatment Center." Optimistically, it is also noted that "she will use strange toilets and tries not to be afraid of dogs."

Are all childhoods alike? In these dry and distant medical transcripts it's possible to see a small child like all other small children and parents like all other parents: anxious, afraid of the slightest abnormality in their adored infant, the father repeatedly assuring the psychologists that Marla is greatly improved, is taking a healthy interest in her dolls, and is making normal sentences like "mother give gun (sic)." "The impression," wrote the Day Center people, "was that the parents were ambivalently fussing over Marla's whining etc., and when they ignored it, it stopped."

Two years later, in 1961, Marla was assessed by a psychiatric clinic. Marla now sat in the interview with a "blank expressionless stare at the ceiling" and refused to tell her name. She spent most of her time, a Dr. Suh notes, "making snakes, big and small, with green clay." Her strong preoccupation with making snakes seemed to swallow up all her energy.

But the recommendation of the doctors is mild. Marla should progress to kindergarten and undergo some psychotherapy. "She still seems to retain," they wrote, "some traces of autistic thinking," but these traces seemed to be regarded as something that could pass, as long as they were not unduly emphasized. Her abnormalities were not seen as debilitating—to the contrary, they are seen as ephemeral.

Marla herself would say that this was the beginning of her catastrophe—this refusal to see her as radically abnormal and to train her accordingly. Falsely mainstreamed, Marla split off from the human race and descended into her present seclusion. It's a seclusion that, as I could see for myself, was extreme, even by the standards of Asperger's unhappiness. Furthermore, Marla seemed to *wield* her unhappiness almost triumphantly, scything down all the things that she found ridiculous because they were inaccessible.

By her own admission, however, this state of affairs might have been avoided if her illness had been caught early enough, which is to say if she had been acculturated in a different way.

"Everything came too late, because I was normalized." Marla was now tapping her knee ferociously and looking tensely at the blinds, as if she expected something appalling, like light, to suddenly erupt through them. "If I hadn't been no normalized, I wouldn't have ended up so abnormal!"

This could well have been true. Again, I could not help thinking of the latahs in their wretched jungle houses: yet they were happy. The latahs were not isolated, their unhappiness (which was sometimes striking) was fitted into a large jigsaw puzzle in which happiness and unhappiness seemed to even themselves out. I thought of the old English adage that our schoolteachers used to brandish at us when they sensed we were becoming churlishly depressed: "be not idle, be not lonely." Was Marla's unhappiness, then, merely the manifestation of her core Asperger's identity, or was it, as she herself suggested, the result of the failures of others? It was very difficult to say.

But at the very least her situation suggested that medicine itself is not enough to treat mental illnesses. Sitting in her apartment, looking at her jars of carob powder, I couldn't help wondering whether the root of Marla's misery was not surprisingly simple, namely a failure to find work. Normality: when all is said and done, is this state nothing more than the holding-down of routine work?

Marla was one end of the spectrum of Asperger's solitude, but at the other was another Montreal resident of about the same age who had turned out very differently. His name was Georges Huard and he was a successful computer programmer at the University of Montreal, as well as being something of a celebrity in French-speaking Canada for his unusual obsession with clocks and time. Almost symbolically, he lived at the other end of town; whereas Marla was despairing and bitter, Georges appeared to be cheery, chirpy, and indefatigably optimistic.

"Oh, Georges Huard," Marla shrugged when I brought up his name on my way out. "He's *much* less Asperger than me."

"Is he?"

"*Much* less. Georges can function."

Marla sounded a little miffed.

"Ask him how happy he is!" she said.

Marla hovered timidly at her door, peering into the hostile, alien corridor into which I was disappearing, as if anxious to close the door behind her once again. I wondered if she was actually jealous of this Georges Huard for his ability to function, his unexpected talent for happiness in an unfriendly world. It didn't seem very likely. Marla merely wanted to leave Montreal for a vacation, but would never leave Montreal for a vacation. And leaving Montreal for a vacation would be happiness for her—it would be like being able to function, whatever that hated word meant.

★

The next day I went to meet happy Georges. The university's computing center is a kitschy oval, sick-yellow building near Métro Saint-Laurent, not far from the housing projects on Maisonneuve where Georges had spent his largely unhappy childhood. Since I had an hour to kill, I wandered around up the hill behind the faculty toward Rue Sherbrooke—Georges had told me in an e-mail that he loved this neighborhood's "*bric-a-brac*," as if it complemented his own personality. Did he mean the Brocantes Baleze of the Rue de Bleury with their dollhouse Eiffel Towers and fig-leaf ashtrays, chairs nailed surreally to their walls, or the bombed-out facades of La Porte d'Arie, with their little bucket stores with names like Africa Bastik, Osaka, and Buttes Impériales, and graffitied Viet cafés? This rolling *quartier* was presumably where Georges walked every day, and at once I felt a kinship with it myself: warehouses, offices, apartment blocks with murals, a Centre de Liquidation de Tapis, a cigar store called Blatter and Blatter on Rue President Kennedy where two frozen men smoked Havanas in the window. The whole *quartier* was a Brocantes Baleze.

As I tramped around, I wondered why Marla hated Montreal so much and why Georges, conversely, loved it so much. Personal idiosyncrasies? Perhaps. But it seemed to me that Marla disliked Montreal because it didn't fit in with her narrow needs, because it was inconvenient. Whereas Georges loved it precisely because it had nothing to do with his needs, because it took him away from the claustrophobia of his needs. His mental world was not constructed out of aversions.

However, hatreds are not less normal than loves, and they certainly didn't in themselves make Marla more abnormal than Georges. We all have hatreds. I have my own, for that matter, things that I detest with a blind passion and that are, in fact, startlingly numerous: instant coffee, for example, or American yogurt, Nutri-Bran bread and *The Catcher in the Rye*. I hate all of these things with an absurd, nitric intensity that is out of all proportion to their actual importance in the world. Anyone of them is capable of putting me into a bad mood for hours on end, especially *The Catcher in the Rye*. I may as well add that my loathing for *The Catcher in the Rye* extends to its style, its tone, and what I regard as its ridiculously undue influence. In truth, I hate *The Catcher in the Rye* every bit as much as Marla hates the lack of Bazooka gum specials in her local supermarket and the Judaic rituals that interfere with her timetables, and I would willingly, if I were dictator, burn every copy of *The Catcher in the Rye* for the greater good of humanity. But what of it? Our hatreds are not just "perseverations," they are also a defense of our personality. Which was why I had not taken Marla's hatreds amiss; they seemed to me a way of defending her personality, hardening it into a definite shape. The only question now was whether Georges, being Marla's antithesis, would be blandly benevolent and affirmative, or whether his enthusiasms would be as strongly defining as my preposterous hatred for *The Catcher in the Rye*.

A French magazine called *Dernière Heure* had given me some idea of what to expect. In a section rather sensationally named *Insolite* (offbeat, strange), Georges was shown sitting by a computer, a hippie with a big grin introduced by a gorgeous headline:

> *On m'appelait Le Savant Fou*
> Je m'endors grâce à un enregistrement des vagues de la mer
> et au "bip" d'un réveil qui sonne tous les 10 minutes.
>
> [*They called me the Mad Genius*
> I can sleep thanks to a sea-waves recording
> and the "beep" of an alarm clock that goes off every ten minutes.]

"Georges Huard," the piece began, "est obsédé par le temps. . . ."

Into the cafeteria of the computing building came a very strange-looking man. The first thing I noticed about Georges was his hair. Like the mane of an aging rock star, it tumbled down in a frayed mass to his waist. But the goggly glasses and serious moustaches immediately switched the first impression to one of professorial eccentricity. Then there was the gait. Georges rolls along on femininely plump hips with one hand dangling loose at the wrist—"my queeny walk," he calls it. In this queenly posture, he sweeps along the curving corridors, his pockets bulging with portable calculators, pagers, and various computing devices. This morning the pager hung around his neck, the Cassiopeia palm pilot filled his breast pocket, and a large money pouch slung around his waist seemed to be filled with yet more tools for making instant calculations. Georges never leaves his house without at least two calculation devices, one of which measures the days, hours, or seconds remaining to a particular event of importance (his birthday or the date of a concert), and the other makes the same calculation backward toward equally significant events (for example, the day on which he began his current job). He appeared fully armed for some kind of war with time.

After introducing me to the autistic kitchen staff of the faculty, who were sitting around a table and looked at Georges with friendly awe, he invited me up to his office on an upper floor. We got into an elevator and Georges fumbled with the big illuminated plastic buttons, all the while explaining the mechanics of his "faggot walk."

"We hold our wrists up like this. Of course, being atypical, we do not stop to consider if this is considered faggoty or not. *Mais non!* We are individualists."

Suddenly, he reminded me of the gay latahs of Kuching: the same assertion of difference, the same dangling wrists, the same odd but unaffected melodrama of appearance. But Georges is not gay; his manner is an effect of his disorder, or so he would say. I had to admit, however, that I had not yet met a single Asperger's man with any of these distinctive traits. Nor had I seen anyone quite so dripping with keys, pouches, and minicomputers. Georges appeared to be *sui generis*.

In any case, we were soon disoriented. Having exited into the floor chosen by Georges, we discovered that it was not the floor we wanted after all. This did not seem to bother Georges too much, and he continued talking:

"I feel disorientated without my Psion computer. I take it with me wherever I go. I feel lost if I don't have it."

"Are you lost now?" I asked, as we floated past scores of identical offices filled with academic secretaries. *Salut, Georges*, they cooed.

"I'm always lost. Especially in this building. People think I'm on drugs, that I'm stoned all the time. Every floor looks like every other floor. Do you mind being lost?"

"Not really. I'm used to it."

"Good man, good man."

Georges was pleased. "Being lost isn't so bad, *n'est-ce pas?*"

We returned to the elevator and fiddled again with the big plastic buttons. They appeared to be a puzzle to Georges, who must have used them almost every day of his life. "Hmm," he mused. "This one?"

What seemed like forty minutes later, we arrived at his office in the Meteorology Department, *Direction des Cycles Supérieurs*. The room was utterly chaotic, with papers, books, and magazines piled high on the desks. These included *Nabuchodinosaure* comic books and copies of the French tabloid *Allo Vedettes*. A plastic pig danced on the computer, as if reveling in the unholy mess of cables, discs, files, and pencils. Here Georges works on computer models of weather systems—a subject that gives him a particular pleasure. He has been obsessed with the weather since the days when he was a humble bike messenger. He showed me some montages he had put together of the recent anti-NAFTA riots in Quebec, then acidly commented that he "hated anarchists of any kind." He told me that he used to be obsessed with insects, that insects more or less dominated his imagination when he was a child, but subsequently stopwatches had taken over. I noticed that he was now staring at my watch rather than looking me in the eye, and it was not long before he was discussing the pros and cons of a dazzling number of different watches.

"That's a Rolex, I see," he said approvingly. "Spent a lot of money on that, eh? I call them 'oysters.' Of course, it doesn't have a countdown timer. The Timex Marathon and the Adidas Runner both have countdown timers. I'm always counting down to something."

I asked him to give an example. "Well," he shot back, "I'm counting down the 6,234,626 seconds to the end of my job contract."

George's sense of time is very microscopic. He counts everything in seconds, including his own age. He'll say, "I was born one billion three hundred thousand seconds ago," and will readily give his age in seconds with constant updates. Since the average human life lasts about 2 billion seconds, he can easily estimate someone's age in years from a second count. On his computer, there were several countdown timers ticking merrily away: the contract, 72 days; an autism conference in Paris, 149 days; and a reminder that in eleven days it would be April 27th. What, I asked, was the significance of those eleven days?

"It means that *statistically* nothing is likely to happen to me over a period of eleven days. It's a reassurance."

Georges reached out and fingered an antique Lab-Chron timer on his desk that, he said, was one of his treasured possessions—it was the first mechanical device able to count seconds precisely.

"There's something beautiful about the flow of seconds. It makes far more sense to me than trying to read the dial of a clock."

Georges's life has been a difficult voyage. For years, his social life consisted of little more than walking endlessly up and down Montreal's Rue Sainte Catherine and reading computer magazines. His mother worked for IBM, and he suspects that she also suffers from Asperger's; his father disappeared when he was five. There were two outstanding events in his youth. On July 10, 1976, when he was seventeen, his mother told him that long hair was now out of style. Georges was traumatized. He suddenly became obsessed with his flowing locks, especially after his math teacher reassured him that he didn't have to cut them.

"Why did hair styles have to change? Why couldn't the rules stay the same? Just because the calendar says 1976—I can't see the point of change at all. . . ."

So Georges kept his hair frozen in the year 1976 simply to avoid a change. For him, it was a matter of spiritual honor. "Sometimes," he added with visible rue, "I'll find myself thinking: Ah, the good old days, when I used to have long hair!"

He tossed his great mane with an expert femininity and tittered. "I think it actually used to be longer."

The other significant event of this deranged youth was his first computer class in 1977. When he mastered his first countdown program, Georges found that at last he could actually do something: he could master time. For the Mindblind, mastering time, he says, "is like sex for you neurotypicals. Time is like sex for me. I obsess about it constantly; it fills my being completely. It gives me my most fundamental pleasure in life." Just as neurotypicals have their various sexual preferences, so Georges prefers one hundred minute units of time—a secret proclivity that he first discovered with his mother's kitchen timer.

This fascination with manipulating time propelled Georges through Montreal's Dawson College and a degree in computing science. But afterward, there followed the by-now-familiar saga of Asperger's calamity: odd jobs, firings, a gradual downward spiral to the edges of mental breakdown, and social leperdom. By his fortieth year, Georges found (somewhat to his horrified surprise) that he was an impoverished aging mailman with an absurd obsession with the weather. At the same time, this deterioration was framed by a methodical observation of rules and laws that bordered on the compulsive. Georges theorizes that Hasidic Jews have distinctly Asperger's characteristics with their regime of 400 commandments and petty rules, not to mention their incessant rocking; was it surprising therefore that Asperger's people were once considered holy? But then again, a passing resemblance to Hasidim and holy men had not served Georges especially well. He was on the brink of disintegration and what many would have called open madness. Without the trust fund enjoyed by someone like Marla, his prospects were grim indeed and getting grimmer by the day.

"Therapy?" I offered weakly.

"I tried it. No good. Especially the psychotherapy—complete crap."

"So what saved you?"

"A man called Peter Zwack. Peter is the guy who hires people here in the Meteorology Department. He has an autistic son, so he knows all the symptoms. By complete chance, I stumbled upon Peter Zwack, and Peter offered me a job. But it wasn't just any old job."

Here the comparison with Marla ended. Zwack decided that Georges could be rebuilt, in a sense, by forming the work he did around his difficult and intractable personality and not the other way around. After all, using play and work as a form of treatment was exactly what Hans Asperger himself had tried to do with his *Heilpädagogik*. It is hard to imagine: half a century ago—in Vienna, during the World War II!—this doctor and his colleagues focusing their cultured, learned minds on the minds of these utterly strange little boys and knowing that it was the fatal solitude of the afflicted individual that would continually deepen his malaise.

It's intriguing to wonder how much mental disorder really is intractable, when the complexity of work is taken into consideration.

The work that Georges is allowed to do has turned his personality around a new axis. He has never been forced to conform to rigid customs whose purpose is normally to make hierarchies both evident and palatable. No one cares if his desk is cluttered with obsessional detritus; his personal manners are understood in advance for what they are, rather than as pathetic *faux pas* in the endless games of office politics. The disadvantages, in other words, of his more egregious abnormalities are obviated by a climate of carefully trained tolerance. His computing skills, after all, need little supervision.

For lunch, I took Georges and Peter Zwack to the restaurant in the Musée de l'Art Contemporain. On the way, we passed through the Place des Arts underground complex, and Zwack, a bluntly energetic man who exuded a paternal protectiveness toward the blinking, long-haired Georges and his rolling "queeny" gait, told me that the office had not had to make any significant concessions to the latter's oddities.

"Very few, in fact. It's simply a question of not taking his abnormality seriously in itself. People can be driven into total disarray or rescued from it. Sometimes it's as simple as having a job."

The museum restaurant is a rotunda held up by huge green columns and made airy by fountains and glass lamps. Georges certainly seemed childish ordering his meal in these elegant surroundings, but his childishness was really the effusion of exuberant glee rather than of autistic social ineptitude. I reflected sadly on how difficult it would have been to ask Marla out to a restaurant, or how different such an occasion would have been with Arthur Ringwalt, who could not enjoy food. Each Asperger's person, in the end, seemed to have a distinct center of gravity and an equally distinct conception of the pleasures of the body. In his 1896 Lowell Lectures, William James observed that "individuals are types of themselves and enslavement to conventional names and their associations is only too apt to blind the student to the facts before him." And he continued:

> The purely symptomatic forms of our classifications are based on the expressive appearances that insanity assumes according to the temper and pattern of the subject whom it affects. In short, individual subjects operate like so many lenses, each one of which refracts in a different angular direction one and the same ray of light.

The same ray of light. I supposed by now that I had seen this "ray" in many of the Asperger's people I had visited or come provisionally to know, for all the varying "refractions" they created. I certainly saw it now in the face of Georges Huard as he ploughed his way through a strawberry dessert, discreetly petting his hand-held Psion calculator and scrutinizing the black dial of my Rolex. Passion, I thought, is passion, whatever its bizarre manifestations and its unfamiliar guises.

"You know," said Peter, "I believe that universities harbor a great number of Asperger's geniuses. One of my mentors was J. M. Marshall, who now has an observatory named after him. He would shout and say inappropriate things—many of his students had nervous breakdowns. But now that I look back at him I think it's obvious that he was Asperger's. I'll go further: I think science departments are filled with them. You can tell; they look like they're lost on a desert island!"

Peter looked queerly triumphant, and Georges nodded vigorously in agreement. Then Peter said:

"Say, have you heard of that guy in New York who is obsessed with subway announcements? He repeats them out of context whenever he feels like it. He'll suddenly blurt out, 'Attention, attention, a train from Manhattan is arriving on platform two. . . .'"

"You mean Darius McCollum?" I said.

"Not sure about the name. But he spends his life trying to get into the subway system and steal trains."

"That's typical behavior," said Georges. "Only more extreme."

"It seems he's quite famous in New York." Peter turned to me. "But you're from New York. Have you heard of this guy?"

"I've heard of him," I said. "I think that would be Darius McCollum."

"I wonder if he has a job?" Georges asked.

"Trains," Peter mused, as if a contemplation of trains might yield a difficult insight pertaining to the inner world of the Asperger's man. "I wonder what it is about the subway?"

Darius McCollum, I thought. For some reason, I had forgotten all about Darius McCollum.

"It's like me with insects," Georges then offered, suddenly looking exquisitely childish and wide-eyed. "It's a world within a world."

"Yes," I said, "that must be what it is."

We walked back out to the Place des Arts and I felt curiously elated to be underground, close to the world of subways and metros. As we shook hands, Peter said to me, "I'd like to meet that Darius McCollum. You really think he steals trains?"

"Assuredly. I'd like to meet him, too."

"Then maybe you will," said Georges. "I'll wager he has an abnormal interest in timetables, too."

He looked down swiftly at his Psion and added, almost in the same breath: "Well, there it is. Time to go. Only 14,445 seconds left till the end of the working day. It's ticking away, isn't it? I didn't think we'd spend 4258 seconds for lunch, but there it is. *Au revoir et bonne chance!*"

★

Over the next few months, I returned to searching the New York City subway for evidence of Darius McCollum. Of course, I knew by now that he was probably safely in jail, but there was always the possibility that he had unexpectedly been released or that he had—would it not be typical?—jumped bail and gone back to his picturesque habits. His trial had been held in March 2001 and been quite a spectacle, fleetingly reported in the *New York Times* as well as in the august *Railway Digest*. He had been defended by the fiery, black civil-rights attorney Stephen L. Jackson, a well-known political figure in Eastern Queens, New York (a middle-class black neighborhood bordering prosperous Nassau County), and supported as well by retired Justice Alvin Schlesinger— all to no avail. Reported the *Times*:

> A Manhattan judge, suggesting she was prepared to be vilified as "the Wicked Witch of the West," sentenced a Queens man yesterday to two and a half years in prison for crimes related to his impersonation of a New York City Transit employee. The judge, Carol Berkman of State Supreme Court in Manhattan, sentenced the man, Darius McCollum, despite letters and other appeals requesting that Mr. McCollum first be assessed by a psychologist who specializes in a neurological disorder, Asperger's syndrome. Asperger's, generally considered a variant of autism, is sometimes marked by a consuming fascination with things like trains. Some people familiar with the condition say Mr. McCollum, who over two decades has been arrested for everything from operating a subway train to driving a city bus, exhibits telltale symptoms.

The proceedings were apparently stormy, with the Wicked Witch of the West sternly lecturing the assembled pleaders.

> The judge, rejecting the idea of Asperger's as a defense, said that Mr. McCollum's family and friends had invoked mental illness as an excuse for his many transit-related crimes and had trivialized the danger he posed to riders. "He could stop doing this if his family and friends would stop telling him, 'Isn't this amusing?'" Justice Berkman said.

"I am shocked," said Justice Schlesinger, after having been reprimanded for second-guessing his colleague on the Manhattan bench. "She tried to make herself the expert in a very complicated field. The only thing I wanted was an adjournment to see if we could help this man. I am disappointed it happened this way."

In issuing her sentence, however, Justice Berkman explained that she had read about Asperger Syndrome on the Internet and had concluded that Darius McCollum did not exhibit the syndrome's most crucial symptoms, including social dysfunction. "As evidence," read the report,

> she pointed to testimony that Mr. McCollum had many friends and was engaged to be married. She also dismissed the notion that Mr. McCollum was not capable of controlling his impulses, which is a common symptom of Asperger's. . . . She said she was mystified by requests that she be lenient with Mr. McCollum, adding that New Yorkers would not want their children to "get on a train driven by Darius McCollum. I have compassion for the mentally ill," the judge concluded fiercely. "Most of the mentally ill lead law-abiding lives. Darius McCollum does not."

As these sad proceedings wound down and the sentence became final, a pitiful scene unfolded in the courtroom:

> Mr. McCollum's mother, Elizabeth, wept quietly as the judge spoke, repeating to herself, "She doesn't know anything about his life." Mr. McCollum, dressed in a denim jacket, did not speak. He smiled at his mother as he left the courtroom. . . . "It is not over yet," Mr. Jackson said. "The judge is dead wrong about Darius."

Although it was now clear that the judge was not about to let Darius resume his nocturnal antics, I held out to myself the hope that some deal might have been struck between Justice Schlesinger and Justice Berkman behind the scenes all the same. Stranger things have happened, and as Asperger's became better known in the media, a greater leniency might be expected to appear in the decisions of even the harshest judges.

But at the same time I now realized that Darius McCollum's subterranean career must definitively have come to a grinding end, that such stunts would now most probably be punished not with a few weeks on Riker's Island but with instant death or deportation to Uzbekistan or with, indeed, exactly what happened: a sentence of two and a half to five years that put Darius McCollum in a maximum security New York prison, 200 miles upstate. Indeed, those same stunts now seemed so innocent as to beggar belief. Could one really, I asked myself, change one's life's obsession to something else, in the same way that Georges had transferred his mental energy from insects to stopwatches? It was difficult to say. Speaking for myself, it seemed to me that one only had two or three obsessions per lifetime, and that it was difficult to imagine giving up *Iron Chef* for something else, much as I often tried to watch other, much less satisfactory, culinary programs such as *Emeril Live* or *Two Fat Ladies*. For that matter, it was equally difficult to give up walking clockwise around lampposts or staying in motels other than a Red Roof Inn. One's habits die hard, and one's obsessions die even harder than one's habits.

And so I have not stopped thinking about Darius McCollum and his nocturnal escapades. Whether they have acquired a dashing underground character or not, they surely remind me of my own most repressed fantasies. Contemplating his tragic career, I never fail to be startled by the modesty of a lost soul who can explain his behavior with the simple words, "I like the scenery. I like the schedules." Sometimes a cigar is just a cigar (as Freud reminds us), and so perhaps a train is just a train.

NOTES

INTRODUCTION

p. ix   "a devastating handicap ..." Uta Frith, "Asperger and His Syndrome," in Frith, 1991, p. 5.

p. ix   "Paucity of empathy; naive, inappropriate, one-sided social interaction ..." Klin, Ami, and Fred Volkmar, "Autism and Asperger's Syndrome," Yale University Child Study Center, July/August 8–9, 1994.

p. xi   "the first plausible variant to crystallize out of the autism spectrum ..." Uta Frith, "Asperger and His Syndrome," in Frith, 1991, p. 5.

p. xi   "More than anything, ADD represents ..." DeGrandpre, 2000, p. 39.

p. xiii   The exchange between Diller and Greene can be read in the archives of *Salon* magazine (http://archive.salon.com). The fireworks began with Diller's review of Greene's *The Explosive Child* on July 18, 2001, and Greene's response on July 19.

p. xvi   "The path to understanding ..." Hans Asperger as quoted in Klin *et al*, 2000, p. xii.

CHAPTER 1: ASPERGER AND I

p. 2   "I like the scenery. I like the schedules." This and other quotations in this chapter by and about Darius McCollum are from "Irresistible Lure of Subways Keeps Landing Impostor in Jail," by Dean E. Murphy, *The New York Times*, August 24, 2000, p. A1.

p. 5   "Autism as a subject touches on the deepest questions of ontology ..." Sacks, 1996, p.246.

p. 6   "Almost all of my social contacts ..." Sacks, 1996, p. 261.

p. 8   "America is the first society to be totally dominated ..." Postman, 1994, p. 145.

p. 9   "What, after all, is normality?" Uta Frith, "Asperger and His Syndrome," in Frith, 1991, p. 23.

p. 10   "There is no reason to suppose that behavior shading into normality ..." Uta Frith, "Asperger and His Syndrome," in Frith, 1991, p. 31.

p. 10   "building a classification system ..." Peter Szatmari, "Perspectives on the Classification of Asperger Syndrome," in Klin *et al*, 2000, p. 408.

p. 11   "The so-called psychiatric specialists ... are the real demons of our age ..." Bernhard, 1988, p. 7.

p. 11   "[R]ules of evidence ..." Peter Szatmari, "Perspectives on the Classification of Asperger Syndrome," in Klin *et al*, 2000, p. 409.

p. 12   "Nevertheless, clinicians feel that ..." Peter Szatmari, "Perspectives on the Classification of Asperger Syndrome," in Klin *et al*, 2000, p. 409.

p. 12   "The DSM-IV lists five disorders ..." All diagnostic descriptions are from The American Psychiatric Association, 2000, pp. 69–84.

p. 21   "A profound lack of affective contact ..." Quoted in Lorna Wing, "The Relationship Between Asperger's Syndrome and Kanner's Autism," in Frith, 1991, pp. 93–94.

p. 21   "Children who do not talk or who parrot speech ..." Uta Frith, "Asperger and His Syndrome," in Frith, 1991, pp. 11–12.

p. 22   "schizoid personality in childhood ..." Sula Wolff, "Schizoid Personality in Childhood and Asperger Syndrome," in Frith, 1991, p. 299.

p. 22   "In his 1944 paper, Asperger gives us the case histories of four boys ..." Quotations are from Hans Asperger's 1944 report as it was reprinted in his textbook *Heilpädagogik* (Berlin, Germany: Springer-Verlag, 1952), as translated by Uta Frith under the title "'Autistic Psychopathy' in Child-hood," in Frith, 1991, pp. 37–92.

p. 29   "Once you catch on to what this syndrome is all about ..." Hallowell, 1995, p. 3.

p. 30   "Your cooperation will be appreciated. A pianist's hands ..." Girard, 1993.

p. 31   "My name is Robert Edwards ..." Fred R. Volkmar, Ami Klin, Robert T. Schultz, Emily Rubin, and Richard Bronen, "Asperger's Disorder," in *American Journal of Psychiatry*, 157, 2000, pp. 262–267.

CHAPTER 2: LITTLE PROFESSORS

p. 47   "Well I was at The Autism Symposium in St Louis again ..." From "The Autism Symposium in St Louis 1997," by David Miedzianik. © 1997 by David Miedzianik. Reprinted by permission. Available at: http://freespace.virgin.net/david.mied/poems/Poem970408.htm.

p. 49   "an increase of this magnitude ..." and following quotations in the paragraph: Diller, 1998, pp. 2–7.

p. 49   Julie Magno Zito; Daniel J. Safer; Susan dosReis; James F. Gardner, ScM; Myde Boles; Frances Lynch, "Trends in the Prescribing of Psychotropic Medications to Preschoolers," *Journal of the American Medical Association*, 283, 2000, pp. 1025–1030.

p. 50   "Is there still a place ..." Diller, 1998, p. 10.

p. 59   "Aloof Group ..." Lorna Wing's four groups are described in her essay "Asperger's Syndrome and Kanner's Autism," in Frith, 1991, p. 109.

p. 62   "Star/big, bright ..." All quotations are from Nicky Werner's privately printed poems, *Thoughts*.

p. 82   "... With what ghoulish glee ..." Miller, 1962, p. viii.

p. 83   "To the anabasis of youth ..." Miller, 1962, p. 155.

CHAPTER 3: THE LAST PURITAN

p. 85   "People are about as important to me as food ..." Payzant, 1978, p. 56.

p. 88   "The history of medicine is full of interesting stories ..." Dr. Harry Angelman's 1965 recollection is reprinted at the website of the Angel-man Syndrome Foundation: http://www.angelman.org/factsofas.htm.

p. 91   "You know/I am deeply in love with a certain beaut. girl ..." Glenn Gould as quoted in Ostwald, 1997, p. 278.

p. 98   "He was reasonably lusty ..." Russell Herbert Gould as quoted in Ostwald, 1997, pp. 40–41.

p. 99   "Had he been autistic, the remarkable success he had ..." Ostwald, 1997, p. 42.

p. 100   "I am delighted to hear that Dr. Gould's perscriptions [sic] ..." Glenn Gould as quoted in Ostwald, 1997, p. 29.

p. 101   "Florence Gould was a woman of propriety ..." Robert Fulford as quoted in Ostwald, 1997, pp. 59–60.

p. 102   "I discovered that, in the privacy ..." Glenn Gould as quoted in Ostwald, 1997, p. 92.

p. 102   "The greatest teacher is the tape recorder ..." Glenn Gould as quoted in Ostwald, 1997, p. 89.

p. 103   "He'd strike off on the bicycle ..." Russell Herbert Gould as quoted in Ostwald, 1997, p. 93.

p. 103   "All eyes turned on the wretched child ..." Pierre Berton as quoted in Ostwald, 1997, p. 62.

p. 103   "I am a skunk, a skunk I am ..." Glenn Gould as quoted in Ostwald, 1997, p. 54.

p. 104   "I wouldn't have, as a child, any toy that was colored red at all ..." Glenn Gould as quoted in Ostwald, 1997, p. 47.

p. 104   "I could imagine what I was doing ..." Glenn Glould as quoted in Ostwald, 1997, p. 77.

p. 104   "What had happened was that the masking noise of the vacuum cleaner ..." Ostwald, 1997, p. 77.

p. 104   "The strange thing was that all of it suddenly sounded better ..." Glenn Glould as quoted in Ostwald, 1997, p. 77.

p. 105   "As we drove there, I noticed a peculiar quality of detachment and isolation ..." Ostwald, 1997, p. 32.

p. 105   "comparable to sitting on the IRT during rush hour ..." Robert Hurwitz as quoted in Payzant, 1978, p. 129.

p. 106   "It really is, in fact, composition ..." Glenn Gould as quoted in Payzant, 1978, p. 130.

p. 106   "... my most joyous moments in radio ..." Glenn Gould as quoted in Ostwald, 1997, p. 259.

p. 107   "G.G.: May I speak now?" This imaginary dialog is quoted in Ostwald, 1997, p. 263.

p. 108   "His current and rather complex diagnosis ..." Helen Metaros, in a transcript of a 1999 panel discussion entitled "Glenn Gould and the Doctors," in *GlennGould Magazine*, vol. 6, no. 2, Fall 2000, p. 88.

p. 108   "Comprising Gould's narcissistic traits ..." Lynn Walter, in a transcript of a 1999 panel discussion entitled "Glenn Gould and the Doctors," in *GlennGould Magazine*, vol. 6, no. 2, Fall 2000. p. 89.

CHAPTER 4: RAIN MEN

p. 115 "Prime numbers and presidents. Petra had never dreamed that she would find these subjects sexy until she met and fell in love with Arthur ..." From "Guess Who Isn't Coming to Lunch," by Jonathan Mitchell. © 2002 by Jonathan Mitchell. Reprinted by permission. Available at: http://hometown.aol.com/jmitch955.

p. 115 "Did I ever tell you that the license plate number ..." From "Guess Who Isn't Coming to Lunch," by Jonathan Mitchell. © 2002 by Jonathan Mitchell. Reprinted by permission. Available at: http://hometown.aol.com/jmitch955.

p. 117 "I'm 'Enery the Eighth I am ..." The ancient music hall lyric (by Murray and Weston, 1911) was first popularized by the British vaudeville singer Harry Champion. In the 1960s, it was re-popularized by the British group Herman's Hermits.

p. 119 "The unconscious which is within us ..." Alfred Binet as quoted in Treffert, 1989, p. 27.

p. 119 "When a mathematician becomes really skilled ..." Grandin, 1996, pp. 32–33.

p. 120 "It is Pullen who comes to mind ..." Treffert, 1989, pp. 32–33.

p. 121 "His powers of observation, comparison, attention ..." Alfred F. Tredgold as quoted in Treffert, 1989, p. 35.

p. 121 "with both his eyes wide open to the bright world ..." F. Sano as quoted in Treffert, 1989, p. 35.

p. 122 "Blind Tom will play this or that piece for you ..." Edward Sequin as quoted in Treffert, 1989, pp. 39–40.

p. 122 "As soon as the new tune begins ..." Edward Sequin as quoted in Treffert, 1989, p. 40.

p. 122 "kicking, pounding his hands together ..." Edward Podolsky as quoted in Treffert, 1989, p. 41.

p. 123 "Most aments are fond of music ..." Alfred F. Tredgold as quoted in Treffert, 1989, p. 41.

p. 123 "The association of musical ability ..." Treffert, 1989, p. 41.

p. 123 "the name, age, address, family structure ..." David S. Viscott as quoted in Treffert, 1989, pp. 45–46.

p. 124 "He could also describe the highways ..." Peek, 1996.

CHAPTER 5: DIAGNOSING JEFFERSON

p. 140 "My Jefferson study reveals a brilliant and talented man ..." Ledgin, 2000, p. 196.

p. 140 "To some his body language appeared odd and awkward ..." Ledgin, 2000, p. 1.

p. 140 "a marked impairment in the use of nonverbal behaviors ..." Ledgin, 2000, p. 8, quoting American Psychiatric Association, 2000, p. 84.

p. 141 "Jefferson's dedication to the Monticello project ..." Ledgin, 2000, p. 87.

p. 142 "a prolonged childhood" Ledgin, 2000, p. 95.

p. 143   "To be a highbrow is the natural state ..." From W. H. Auden, "Letter to Lord Byron," in *Collected Poems*, Vintage paperback ed. 1991, p. 101.

p. 143   "diplomacy of the interior regions" Ellis, 1996, p. 105.

p. 143   "In other cases, it was an orchestration of his internal voices ..." Ellis, 1996, p. 106.

p. 143   "He found dining with Jefferson exasperating ..." Gore Vidal, "Amistad," in Vidal, 2001, p. 170.

p. 144   "the fictitious principle ..." Thomas Jefferson as quoted in Ellis, 1996, p. 36.

p. 144   "crucial clue to Jefferson's ..." Ellis, 1996, p. 37.

p. 144   "complete fabrication" Ellis, 1996, p. 39.

p. 144   "The explanation lies buried ..." Ellis, 1996, p. 39.

p. 144   "magical mystery tour of architectural legerdemain" John Randolph as quoted in Ellis, 1996, p. 40.

p. 144   "suggested a level of indulged ..." Ellis, 1996, p. 40.

p. 145   "In Jefferson's case, specific-to-general thinking is what made Monticello ..." Temple Grandin in Ledgin, 2000, p. 199.

CHAPTER 6: AUTIBIOGRAPHIES

p. 153   "Andy Horowitz was nine years old ..." From "The Session," by Jonathan Mitchell. © 2002 by Jonathan Mitchell. Reprinted by permission. Available at: http://hometown.aol.com/jmitch955.

p. 155   "Biff then penetrated Monique with his penis ..." From "Blot," by Jonathan Mitchell. © 2002 by Jonathan Mitchell. Reprinted by permission. Available at: http://hometown.aol.com/jmitch955.

p. 156   "Arnold Springer goes to a doctor ..." From "Questionmark Etiology," by Jonathan Mitchell. © 2002 by Jonathan Mitchell. Reprinted by permission. Available at: http://hometown.aol.com/jmitch955.

p. 157   "Other guys could do well with the chicks doctor ..." Jonathan Mitchell. From "Questionmark Etiology," by Jonathan Mitchell. © 2002 by Jonathan Mitchell. Reprinted by permission. Available at: http://hometown.aol.com/jmitch955.

p. 157   Who you see here ..." From "Questionmark Etiology," by Jonathan Mitchell. © 2002 by Jonathan Mitchell. Reprinted by permission. Available at: http://hometown.aol.com/jmitch955.

p. 159   "Even today, my thinking is from the vantage point of an observer ..." Grandin, 1996, p. 132.

p. 150   "In order to deal with a major change ..." Grandin, 1996, p. 34.

p. 160   "Each door or gate ..." Grandin, 1996, p. 35.

p. 161   "The phone bill came on Friday ..." From "Sunday 21st December 1997," by David C. Miedzianik. © 1997 by David C. Miedzianik. Reprinted by permission. Available at: http://freespace.virgin.net/david.mied/poems/Poem971221.htm.

p. 161   "Well it's New Year's Day 1998 today ..." From "New Years Day 1998," by David C. Miedzianik. © 1998 by David C. Miedzianik. Reprinted

by permission. Available at:
http://freespace.virgin.net/david.mied/poems/Poem980101b.htm.

p. 162   "Why someone won't come to love …" From "My Emotions Are
           Driving Me Crazy," by David C. Miedzianik. © 1997 by David C.
           Miedzianik. Reprinted by permission. Available at:
           http://freespace.virgin.net/david.mied/poems/Poem971228.htm.

p. 162   "There's too much isolation for me in Rotherham …" From "Words on
           David, by Hubert," an introduction to David Miedzianik's poems.
           © 1997 by David C. Miedzianik. Reprinted by permission. Available at:
           http://freespace.virgin.net/david.mied/hubert.htm.

p. 163   "When Temple Grandin finished talking …" From "The Autism
           Symposium in St Louis 1997," by David C. Miedzianik. © 1997 by
           David C. Miedzianik. Reprinted by permission. Available at:
           http://freespace.virgin.net/david.mied/poems/Poem970408.htm.

p. 163   "I can remember going into Derbyshire in my mum's car …" From
           "I Have Trouble Summing My Feelings Up in Words," by David C.
           Miedzianik. © 1997 by David C. Miedzianik. Reprinted by permission.
           Available at:
           http://freespace.virgin.net/david.mied/poems/Poem970804.htm.

p. 164   "My father was a podiatrist …" From the Introduction to "Diathermy,"
           a poem by David Spicer. © 2002 by David Spicer. Reprinted by permis-
           sion. Available at:
           http://www.udel.edu/bkirby/asperger/dave_spicer_poems.html.

p. 164   "When first the ancient switch was thrown …" From "Diathermy," by
           David Spicer. © 2002 by David Spicer. Reprinted by permission.
           Available at:
           http://www.udel.edu/bkirby/asperger/dave_spicer_poems.html.

p. 165   "Something over twenty years ago …" From the Introduction to "Good
           to the Last Drop," a poem by David Spicer. © 2002 by David Spicer.
           Reprinted by permission. Available at:
           http://www.udel.edu/bkirby/asperger/dave_spicer_poems.html.

p. 165   "I tried to make my cohorts green …" From "Good to the Last Drop,"
           by David Spicer. © 2002 by David Spicer. Reprinted by permission.
           Available at:
           http://www.udel.edu/bkirby/asperger/dave_spicer_poems.html.

CHAPTER 7: THE POETICS OF MEDICINE

p. 181   "Viewed in mental and social health perspectives …" Arthur Kleinman,
           "Introduction to the Transformation of Social Experience in Chinese
           Society," in *Culture, Medicine, and Psychiatry*, vol. 23, no. 1, 1999,
           pp. 1–6.

p. 181   "The celebrated historian Eric Hobsbawm …" Arthur Kleinman,
           "Introduction to the Transformation of Social Experience in Chinese
           Society," in *Culture, Medicine, and Psychiatry*, vol. 23, no. 1, 1999, pp. 1–6.

p. 182   "... mental illnesses are real, but ..." Kleinman, 1988, p. 3.

p. 182   Geertz, "Latah in Java: A Theoretical Paradox," in *Psychiatry*, vol. 22, no. 3, 1968, pp. 225–237.

p. 200   "... individuals are types of themselves ..." William James in Eugene Taylor (ed.), *William James on Exceptional Mental States: The 1896 Lowell Lectures*. New York: Charles Scribner's Sons, 1983.

p. 202   "A Manhattan judge, suggesting she was prepared to be vilified ..." This and the following quotes from the trial were reported in "Judge, Clearly Not Amused, Sentences a Subway Impostor" by Dean E. Murphy, *The New York Times*, March 30, 2001, p. B3.

BIBLIOGRAPHY

American Psychiatric Association, *Diagnostic and Statistical Manual of Mental Disorders, Fourth Edition, Text Revision*. Washington, D.C.: The American Psychiatric Association, 2000.

Bernhard, Thomas, *Wittgenstein's Nephew: A Friendship*. New York: Alfred Knopf, 1988.

DeGrandpre, Richard, *Ritalin Nation: Rapid-Fire Culture and the Transformation of Human Consciousness*. New York: W. W. Norton, paperback ed. 2000.

Diller, Lawrence H., *Running on Ritalin: A Physician Reflects on Children, Society, and Performance in a Pill*. New York: Bantam Books, 1998.

Ellis, Joseph, *American Sphinx: The Character of Thomas Jefferson*. New York: Vintage Books, 1998.

Fling, Echo, *Eating an Artichoke: A Mother's Perspective on Asperger Syndrome*. London, U.K.: Jessica Kingsley Publishers, 2000.

Frith, Uta (ed.), *Autism and Asperger Syndrome*. Cambridge, U.K.: Cambridge University Press, paperback ed. 1991.

Girard, Francois, *Thirty-Two Short Films About Glenn Gould*. Toronto, Canada: Rhombus Media, 1993.

Grandin, Temple, *Emergence: Labeled Autistic*. New York: Warner Books, 1986.

Grandin, Temple, *Thinking in Pictures: And Other Reports from My Life with Autism*. New York: Vintage Books, 1996.

Hamilton, Lynn, *Facing Autism: Giving Parents Reasons for Hope and Guidance for Help*. Colorado Springs, Colorado: Waterbrook Press, 2000.

Hallowell, Edward, *Driven to Distraction: Recognizing and Coping with Attention Deficit Disorder from Childhood through Adulthood*. New York: Simon & Schuster, paperback ed. 1995.

Kleinman, Arthur, *Rethinking Psychiatry: From Cultural Category to Personal Experience*. New York: The Free Press, 1988.

Klin, Ami; Fred R. Volkmar; and Sara S. Sparrow (eds.), *Asperger Syndrome*. New York: The Guilford Press, 2000.

Ledgin, Norm, *Diagnosing Jefferson: Evidence of a Condition That Guided His Beliefs, Behavior, and Personal Associations*. Arlington, Texas: Future Horizons, Inc., 2000.

Miller, Henry, *The Time of the Assassins: A Study of Rimbaud*. New York: New Directions Paperbook, 1962.

Ostwald, Peter F., *Glenn Gould: The Ecstasy and Tragedy of Genius*. New York: W. W. Norton, paperback ed. 1997.

Payzant, Geoffrey, *Glenn Gould: Music and Mind*. Toronto, Canada: Van Nostrand Reinhold, 1978.

Peek, Fran, *The Real Rain Man: Kim Peek*. Salt Lake City, Utah: Harkness Publishing Consultants, 1997.

Postman, Neil, *The Disappearance of Childhood*. New York: Vintage Books, 1994.

Simons, Ron and Charles Hughes (eds.), *Culture-Bound Syndromes: Folk Illnesses of Psychiatric and Anthropological Interest*. Dordrecht, Netherlands: D. Reidel Publishing Company, 1985.

Sacks, Oliver, *An Anthropologist on Mars: Seven Paradoxical Tales*. New York: Vintage Books, 1996.

Treffert, Darold A., *Extraordinary People: Understanding Savant Syndrome*. New York: Ballantine Books, 1989.

Vidal, Gore, *The Last Empire: Essays 1992–2000*. New York: Vintage Books, 2001.

Willey, Diane, *Pretending to be Normal: Living with Asperger's Syndrome*. London, U.K.: Jessica Kingsley Publishers, 1999.

ACKNOWLEDGMENTS

The description of the New York City Asperger's school in Chapter 2 originally appeared, in a different form, in *The New York Times Magazine*.

The discussion of culture-bound syndromes and latah in Chapter 7 originally appeared, in a different form, in *The New York Times Magazine*.

The stories of Jonathan Mitchell, in Chapter 6, are copyrighted by the author and used with his permission. (See specific copyright notices under "Notes.")

The poems of David Miedzianik, in Chapter 6, are copyrighted by the author and used with his permission. (See specific copyright notices under "Notes.")

The poems of David Spicer, in Chapter 6, are copyrighted by the author and used with his permission. (See specific copyright notices under "Notes.")

INDEX

The Time of the Assassins (Miller),
    81–83
tornadoes, 54–55
Tourette Syndrome, 171–172
traffic systems, 9, 18, 136–139
trains, stealing, 1–4, 6–7, 201
Treffert, Darold, 118, 120–121, 123
Tregold, Alfred, 119–121, 123
twiddling, 157–158

V
vacuum cleaners, 76–79
Van Gogh, Vincent, 140
Vidal, Gore, 143
Viscott, David, 123
Volkmar, Fred, 29, 60, 75

W
Walter, Lynne, 108–109
website, for Glenn Gould, 109
Werner, Nicky, 61–62, 70–74
West Africa, mental disorders in,
    180
Wilde, Oscar, 106
Willey, Liane, 43
Williams, Donna, 155
Wing, Lorna, 28–29, 59
Wittgenstein, Ludwig, 100
Wittgenstein, Paul, 10–11
Wittgenstein's Nephew (Bernhard),
    10–11
Witzmann, A., 119
Wolff, Sula, 22
women with Asperger's, 5. *See also*
    Conn, Marla; female savants;
    Grandin, Temple
writers, Asperger's or autistic,
    151–175. *See also* Grandin,
    Temple; poets, Asperger's or
    autistic

Z
Zak, Viktorine, 19
Zwack, Peter, 199–200
Zweig, Stefan, 179